MODELLING OF CATHODIC PROTECTION SYSTEMS

WITPRESS

WIT Press publishes leading books in Science and Technology.
Visit our website for the current list of titles.
www.witpress.com

WITeLibrary

Home of the Transactions of the Wessex Institute, the WIT electronic-library provides the
international scientific community with immediate and permanent access to individual
papers presented at WIT conferences. Visit the WIT eLibrary at www.witpress.com

Advances in Boundary Elements Series

Objectives

The continuing interest in the application of the Boundary Element Method has generated a series of books and numerous scientific papers. In spite of all their advantages, the need exists for a serial publication in which the most recent advances in the method are demonstrated in a more complete form.

Each volume in the Series comprises authored or edited books written by leading researchers in the field. The volumes are all self contained and cover a particular topic in sufficient detail for the analyst to understand the subject. Some books report on practical applications of the technique.

The Series covers topics such as:

Fluid Mechanics	Acoustics
Heat Transfer	Cathodic Protection Problems
Stress Analysis	High Performance Computing
Fracture Mechanics	Sparse Methods
Contact Mechanics	Numerical Integration
Structural Dynamics	Industrial Applications
Inelastic Problems	Basic Principles
Optimization and Sensitivity	Electrical and Electromagnetic Problems
Plate Bending	Mathematical & Computational Aspects

Series Editor

C.A. Brebbia
Wessex Institute of Technology,
Ashurst Lodge, Ashurst, Southampton, SO40 7AA, UK
Email: carlos@wessex.ac.uk
Telephone: 44 (0) 238 029 3223
Fax: 44 (0) 238 029 2853

Honorary Editor

M. Jaswon
City University, UK

Associate Editors

V. Leitao
Inst. Superior Tecnico
Portugal

W.J. Mansur
COPPE,
Federal University of Rio de Janeiro
Brazil

R.A. Meric
Research Institute for Basic Sciences
Turkey

K. Onishi
Ibaraki University
Japan

D. Ouazar
Universite Mohammed V
Morocco

F. Paris
University of Seville
Spain

M. Predeleanu
Universite Paris VI
France

J.J. Rencis
University of Arkansas
USA

T.J. Rudolphi
Iowa State University
USA

A.P. Selvadurai
McGill University
Canada

R.P. Shaw
State University of New York
USA

P. Skerget
University of Maribor
Slovenia

V. Sladek
Slovak Academy of Sciences
Slovakia

S. Syngellakis
University of Southampton
UK

M. Tanaka
Shinshu University
Japan

N. Tosaka
Nihon University
Japan

T. Tran-Cong
University of Southern
Queensland
Australia

W.S. Venturini
University of Sao Paulo
Brazil

J.L. Wearing
University of Sheffield
UK

MODELLING OF CATHODIC PROTECTION SYSTEMS

Editor

R.A. Adey
C. M. BEASY Ltd.
Wessex Institute of Technology, UK

WITPRESS Southampton, Boston

MODELLING OF CATHODIC PROTECTION SYSTEMS

Series: Advances in Boundary Elements, Vol. 12

Editor: R.A. Adey

Published by

WIT Press

Ashurst Lodge, Ashurst, Southampton, SO40 7AA, UK
Tel: 44 (0) 238 029 3223; Fax: 44 (0) 238 029 2853
E-Mail: witpress@witpress.com
http://www.witpress.com

For USA, Canada and Mexico

WIT Press

25 Bridge Street, Billerica, MA 01821, USA
Tel: 978 667 5841; Fax: 978 667 7582
E-Mail: infousa@witpress.com
http://www.witpress.com

British Library Cataloguing-in-Publication Data

A Catalogue record for this book is available
from the British Library

ISBN: 1-85312-889-9
ISSN: 1460-1419

Library of Congress Catalog Card Number: 2004101299

Contents

Preface

The field of Corrosion Modelling has evolved tremendously since the first work was published in the early 1980's. Its initial application in the offshore industry has expanded to the point where modelling is applied in practically all application areas and the software has developed to fulfil the needs. This book presents contributions from the most influential researchers and developers of corrosion modelling tools and users who apply the technology in their industry.

Applications are presented from a wide variety of industries including Oil and Gas, Defence, Ship Building, Maritime and Underground Utilities. Specific applications include design and optimisation of active (ICCP) and sacrificial CP systems, modelling performance of corrosion control systems for pipelines, prediction of interference from tramways and other electrical sources on CP systems, prediction and management of corrosion related electric and magnetic fields, inverse modelling to provide data for condition monitoring etc.

This book provides an excellent introduction to the state of the art in computer modelling of corrosion and related electrochemical processes and will be of interest to corrosion engineers and physicists, model developers and researchers.

R.A. Adey
2005

Mathematical modeling for corrosion analysis

K. Amaya

Graduate School of Information Science and Engineering, Department of Mechanical and Environmental Informatics, Tokyo Institute of Technology, O-okayama, Meguro-ku, Tokyo 152, Japan.

1 Introduction

Corrosion is all around us and can affect our lives in many ways, because metals are so widely used in our world. Corrosion-related failures in the industries often result in catastrophic failures. Corrosion is not only dangerous, but costly. The cost of corrosion is estimated to be huge and protecting or reducing corrosion is one of the most important problems in engineering. Many attempts have been made to predict analytically corrosion behaviors [1].

In this chapter, firstly, we will explain basics of corrosion as electro-chemical process. The principle of Voltaic cells is explained, in order to help the understanding about basics of corrosion phenomena at the point of view of electro chemistry. We also refer to the principles of Voltaic cells Galvanic corrosion and cathodic protection.

After we have a basic understanding of the electro-chemical nature of corrosion and behavior of metals in corrosive fluids, we will explain quantitative treatments of corrosion rate. Finally, a mathematical modeling for corrosion rate analysis is constructed.

Purely mathematical treatments would have to be limited to simplified geometric electro-chemical sources. However, thanks to the recent progress of numerical analysis method such as finite element method and boundary element method, corrosion rate analysis of practical structure can be performed by using proposed model [2–4].

2 Basics of corrosion

2.1 Corrosion as an electro-chemical process

Corrosion is an electro-chemical process in which metal atoms are oxidized from positive ions while other chemical species are reduced [6]. This results in a flow of electrons from one site on the metal surface to another. A piece of bare iron left outside where it is exposed to moisture will rust quickly. If the moisture is salt water, it will rust even more quickly. The corrosion is accelerated by an electro-chemical in process oxidizing the iron. A water drop forms a voltaic cell in contact with the metal in this process.

Let's consider a water oxidizing iron supplies electrons at the edge of the water drop to reduce oxygen from the air as shown in fig. 1. At the iron surface inside the water drop an oxidation process occurs. This involves a loss of electrons. A Site where oxidation occurs is defined as Anode.

$$Fe(s) \rightarrow +Fe^{2+}(aq) + 2e^-\qquad(1)$$

For this reaction to occur, simultaneous reduction process, which is a gain of electrons, should occur as the following.

$$O_2(g) + 2H_2O(l) + 4e^- \rightarrow 4OH^-(aq)\qquad(2)$$

The site where reduction takes place is defined as cathode. Cathode action reduces oxygen from air, forming hydroxide. The hydroxide ions can move into the water drop to react with the iron(II) ions moving from the oxidation region. Iron(II) hydroxide forms and precipitates.

$$Fe^{2+}(aq) + 2OH^-(aq) \rightarrow Fe(OH)_2(s)\qquad(3)$$

Producing rust by the oxidation of the precipitate is expressed with the following process:

$$4Fe(OH)_2(s) + O_2(g) \rightarrow 2Fe_2O_3 * H_2O(s) + 2H_2O(l)\qquad(4)$$

The rusting of bare iron in the presence of air and water cannot be avoided because it is driven by an electro-chemical process. However, other electro-chemical processes are applicable to protection against corrosion.

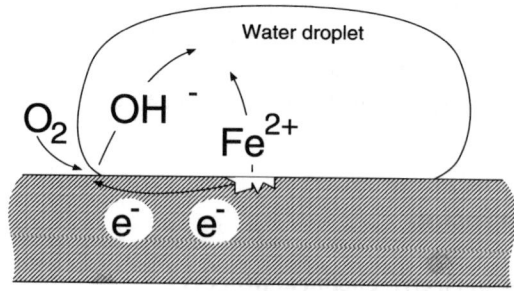

Figure 1: Electro-chemical process of iron corrosion.

2.2 Galvanic corrosion

Galvanic corrosion refers to corrosion damage induced when two different materials are coupled in a corrosive electrolyte [7]. This is also called 'different metal corrosion' or wrongly 'electrolysis'. It occurs when two (or more) different metals are brought into electrical contact under solution such as sea water. When a galvanic couple forms, one of the metals in the couple becomes the anode and corrodes faster than it would by itself. On the other hand, one of the metals becomes the cathode and corrodes slower than it would alone. Both metals in the couple may or may not corrode by themselves in sea water. When contact with a different metal is made, however, the self corrosion rates of both anode and cathode will change into accelerated and decelerated/stopped, respectively.

Galvanic corrosion can be one of the most common forms of corrosion as well as one of the most destructive. The driving force for corrosion is a potential difference between the different materials. Figure 2 shows the concept for help in understanding of driving force of corrosion. In order to analyse the corrosion rate, it is important to quantify this potential difference. In section 3, we will explain the electro chemical treatment which handles this driving force (potential gap) quantitatively.

2.3 Voltaic cells

Consider zinc and copper metals connected with a lead wire and placed in solutions such as H_2SO_4 as shown in fig. 3. This is one of the simplest ways to generate a sustained electrical current. An electro-chemical cell which causes external electric current flow can be created using any two different metals since metals differ in their tendency to lose electrons. Zinc more readily loses electrons than copper, so placing zinc and copper metals in solutions of their salts can cause electrons to flow through an external wire which leads from the zinc to the copper.

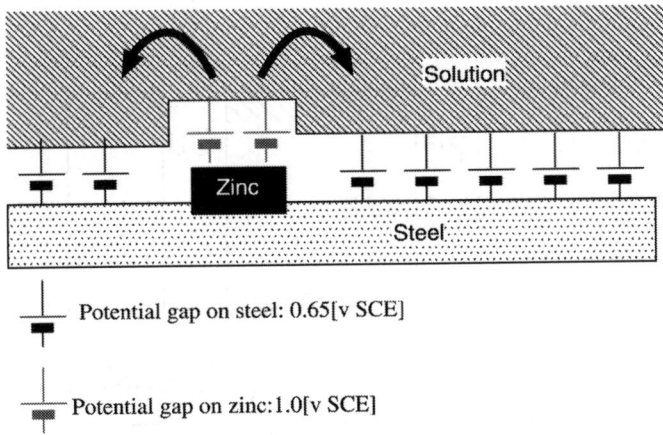

Figure 2: Conceptual diagram for imaging driving force of corrosion.

As a zinc atom provides the electrons, it becomes a positive ion and goes into aqueous solution, decreasing the mass of the zinc electrode.

On the copper side, the two electrons received allow it to convert oxygen from solution into hydroxide. The two reactions are typically written

$$Zn \rightarrow Zn^{2+}(aq) + 2e^- \tag{5}$$

$$H_2 + 2e^- \rightarrow H_2(g) \tag{6}$$

Energy is required to force the electrons to move from the zinc to the copper electrode, and the amount of energy per unit charge available from the voltaic cell is called the electro-motive force (emf) of the cell. Energy per unit charge is expressed in volts (1 volt = 1 joule/coulomb).

Clearly, to get energy from the cell, you must get more energy released from the oxidation of the zinc than it takes to reduce the hydrogen. The cell can yield a finite amount of energy from this process, the process being limited by the amount of material available either in the electrolyte or in the metal electrodes. For example, if there were one mole of the sulfate ions SO_4^{2-}, then the process is limited to transferring two moles of electrons through the external circuit. The amount of electric charge contained in a mole of electrons is called the Faraday constant, and is equal to Avogadro's number times the electron charge:

$$Faraday \; constant = F = N_A e = 6.022 \times 10^{23} \times 1.602 \times 10^{-19}$$

$$= 96,485 \; Coulombs/mole \tag{7}$$

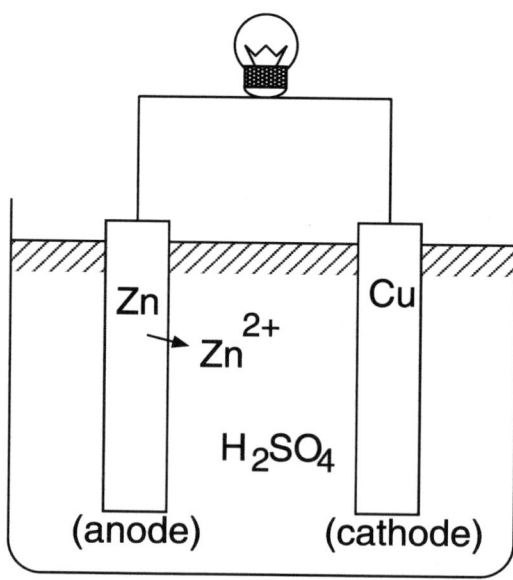

Figure 3: Schematic diagram of Voltaic cell.

The energy yield from a voltaic cell is given by the cell voltage times the number of moles of electrons transferred times the Faraday constant.

$$\text{Electrical energy output} = nFE_{cell} \qquad (8)$$

The energy yield from the standard electrode potentials for both the reactions. For the zinc/hydrogen cell under the standard conditions, the calculated cell potential is 0.76 volts.

2.4 Cathodic protection against corrosion

Cathodic protection operates according to the same principle as voltaic cell [8]. This principle was also engineered into the protection of metallic structures. The sacrificial corrosion of one metal such as zinc, magnesium or aluminum is a well-known method of cathodically protecting metallic structures. In a bimetallic couple, the less noble material will become the anode of this corrosion cell and tend to corrode at an accelerated rate, compared with the uncoupled condition. The more noble material will act as the cathode in the corrosion cell. The flow of cathodic protection current causes the zinc anode to corrode (oxidation), while the steel pipe is protected from corrosion (reduction).

Underground steel pipes offer the strength to transport fluids at high pressures, but they are vulnerable to corrosion driven by electro-chemical processes. A measure of protection can be offered by driving a zinc rod into the ground near the pipe and providing an electrical connection to the pipe. Since the zinc has a standard potential of -0.76 volts compared to -0.41 volts for iron, it can act as an anode of a voltaic cell with the steel pipe acting as the cathode. With soil serving as the electrolyte, a small current can flow in the wire connected to the pipe. The zinc rod will be eventually consumed by the reaction

$$\text{Zn(s)} \rightarrow +\text{Zn}^{2+}(\text{aq}) + 2e^- \qquad (9)$$

while the steel pipe as the cathode will be protected by the reaction

$$\text{O}_2(\text{g}) + 2\text{H}_2\text{O}(\text{l}) + 4e^- \rightarrow 4\text{OH}^-(\text{aq}). \qquad (10)$$

Cathodic protection is also commonly employed by impressed current systems. The electrical circuit is similar to that of a sacrificial anode system except for the fact that a DC power supply is used.

3 Corrosion modeling for corrosion rate analysis

3.1 Standard electrode potentials

In an electro-chemical cell, an electric potential is created between two different metals [9]. This potential is a measure of the energy per unit charge which is available from the oxidation/reduction reactions to drive the reaction.

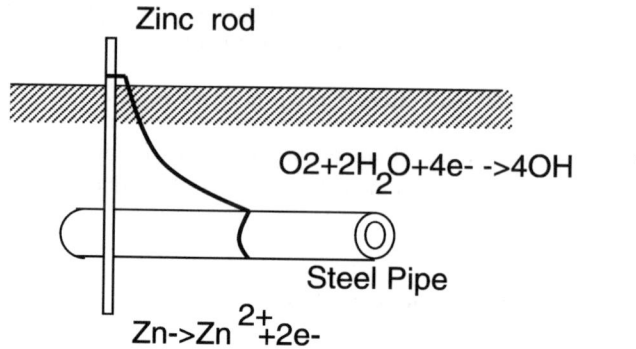

Figure 4: Schematic diagram of cathodic protection.

The driving force for the above reaction can be measured by placing a voltage measuring device in the circuit. We may consider this force as being the sum of two potentials called half cell potentials or single electrode potentials, one of these is associated with the half cell reaction occurring at the anode and the other is associated with the half cell reaction taking place at the cathode.

It is customary to visualize the cell reaction in terms of two half-reactions, an oxidation half-reaction and a reduction half-reaction.

$$\text{Reduced species} \rightarrow \text{oxidized species} + ne^- \quad \text{Oxidation at anode} \tag{11}$$

$$\text{Oxidized species} + ne^- \rightarrow \text{reduced species} \quad \text{Reduction at cathode} \tag{12}$$

The cell potential has a contribution from the anode which is a measure of its ability to lose electrons 'it will be called its oxidation potential'. The cathode has a contribution based on its ability to gain electrons, its 'reduction potential'. The cell potential can then be written

$$E_{cell} = \text{oxidation potential} + \text{reduction potential} \tag{13}$$

If we could tabulate the oxidation and reduction potentials of all available electrodes, then we could predict the cell potentials of voltaic cells created from any pair of electrodes. Actually, tabulating one or the other is enough, since the oxidation potential of a half-reaction is the negative of the reduction potential for the reverse of that reaction.

To obtain consistent relative half cell potential data, it is necessary to compare all electrodes against a common reference. The reference electrode should be easy to construct, exhibit reversible behavior, and give constant and reproducible potentials for a given set of experimental conditions. The standard hydrogen electrode (SHE) meets these requirements and is commonly used as the ultimate reference electrode.

Thus, two main difficulties must be overcome to establish such a tabulation. (1) We cannot measure absolute potentials for half cell reactions but the relative half cell

Table 1: List of standard electrode potentials for common metals.

Metal		Standard Electrode Potential [V]
	Active End	
Potassium	$K \leftrightarrow K^+ + e^-$	$-\,2.92$
Magnesium	$Mg \leftrightarrow Mg^{2+} + 2e^-$	$-\,2.38$
Aluminum	$Al \leftrightarrow Al^{3+} + 3e^-$	$-\,1.66$
Zinc	$Zn \leftrightarrow Zn^{2+} + 2e^-$	$-\,0.76$
Chromium	$Cr \leftrightarrow Cr^{3+} + 3e^-$	$-\,0.71$
Iron	$Fe \leftrightarrow Fe^{2+} + 2e^-$	$-\,0.44$
Nickel	$Ni \leftrightarrow Ni^{2+} + 2e^-$	$-\,0.23$
Hydrogen	$2H^+ + 2e^- \leftrightarrow H_2$	0.00
		Reference
Copper	$Cu \leftrightarrow Cu^{+2} + 2e^-$	$+0.34$
Silver	$Ag \leftrightarrow Ag^+ + e^-$	$+0.80$
Platinum	$Pt \leftrightarrow Pt^{+2} + 2e^-$	$+1.20$
Gold	$Au \leftrightarrow Au^{+3} + 3e^-$	$+1.42$
	Noble or Passive End	

potentials that can be measured are quite useful. (2) The electrode potential depends upon the concentrations of the substances, the temperature, and the pressure in the case of a gas electrode.

In practice, the first of these difficulties is overcome by measuring the potentials with respect to a standard hydrogen electrode.

The electrode basically consists of a platinum wire immersed in a solution containing hydrogen ions, and hydrogen gas is bubbled across the surface of the platinum. This type of electrode is called a gas electrode because the platinum takes no part in the electro-chemical reaction. The half cell reaction of the cell is given as

$$H_2(g) \rightarrow 2H^+ + 2e^- \tag{14}$$

It is the nature of electric potential that the zero of potential is arbitrary; it is the difference in potential which has a practical consequence. Tabulating all electrode potentials with respect to the same standard electrode provides a practical working framework for a wide range of calculations and predictions. The standard hydrogen electrode is assigned a potential of zero volts. The standard electrode potential of commonly encountered metal is shown in table 1.

3.2 Polarization

When one measures the normal electrode potential, it is noted that each half-cell reaction of oxidation and reduction is at equilibrium in a solution of its own ions compared to the potential of the reference electrode. To be equilibrium means that an equal number of ions are always being produced and reduced.

In case this equilibrium is not balanced, the electrode potential starts to drop for cathodic reaction and increase for anodic reaction. This potential shift is called polarization. The equilibrium can be braked by impressing external current to the metal electrode. Polarization is defined as the change in the potential of an electrode from the open circuit potential to the potential resulting from a current flow.

The relationship between the potential change and the net current density is also called a polarization curve which is usually a non-linear function.

A simple experimental setting for measuring a polarization is shown in fig. 5. The variable resistance is used to change the externally impressed current. Platinum is often used for counter electrode, because it will not polarize excessively, in order not to contaminate the solution. The reference electrode is frequently Ag/Ag chloride or Hg/calomel in KCl solution. Experimental setting often employs a Luggin capillary placed very close to the specimen, which enables measuring the potential at a specific place on a specimen surface.

Since the potential is changing refer to the current density through the surface of material, it is important to model this polarization effect for corrosion rate analysis. The conceptual diagram for understanding driving force of corrosion as shown in fig. 2 can be updated as in fig. 6, when the polarization characteristics are modeled.

3.3 Corrosion rate

The current which flows through the surface of metal per unit area is called corrosion current density i_{corr}. Referring to fig. 1, one iron ion is released into the electrolyte

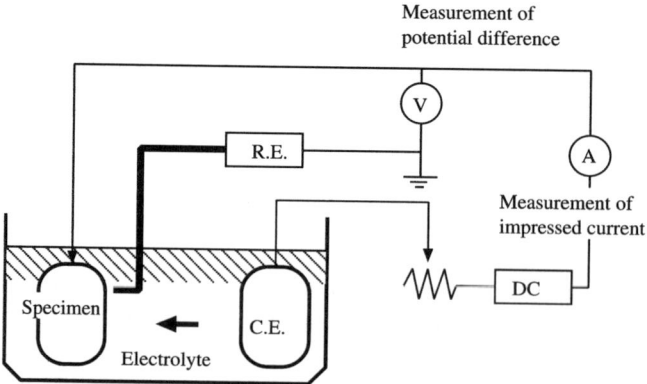

Figure 5: Simple experimental setting for measuring polarization curve.

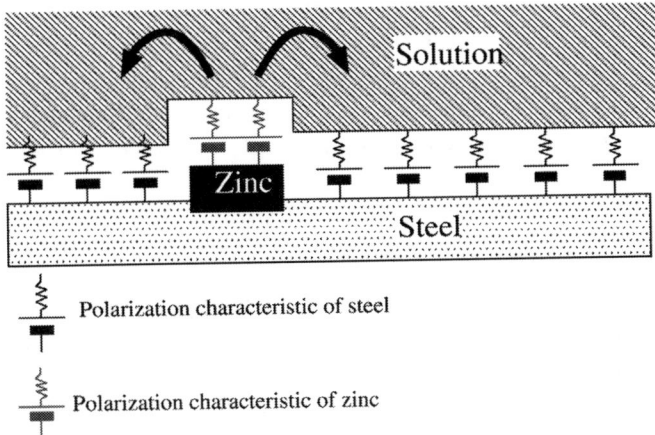

Figure 6: Conceptual diagram for imaging driving force of corrosion.

for each 2 electrons which flow into the metal surface. Because i_{corr} is a measure of electron flow per unit time and unit area, it is obviously proportional to amount of metal loss per unit time. i_{corr} is a good measure of corrosion rate.

Corrosion current [mA/cm^2], mass loss [g/m^2 day] and penetration rates [mm/y or mpy] are often used for express corrosion rates. The following equation provides a simple way to convert data from corrosion current i_{corr}[mA/cm^2] to penetration rates d[mm/yr].

$$d[\text{mm/yr}] = 3.28 i_{corr} M / n\rho[\text{mA/cm}^2] \tag{15}$$

where, [mpy] is milli-inch per year, n is number of electrons freed by the corrosion reaction, M[g] is atomic mass and ρ[g/cm^3] is density. Since we are able to know the relationship between current density and potential by measuring the polarization curve, once one could obtain the potential value on the metal surface, the current density (corrosion rate) can be estimated.

4 Mathematical modeling for corrosion rate analysis

In case of assuming the electro-chemical reaction on the metal surface is not interfered by solution velocity, the current density in the electrolyte domain Ω can be expressed with the following electro migration diffusion equation [5, 10–12]:

$$\boldsymbol{i} = F \sum_j z_j D_j \text{grad} c_j - F^2 \sum_j z_j^2 c_j u_j \text{grad}\phi \tag{16}$$

where $\boldsymbol{i}$ is the current density vector, F is Faraday's constant, z_j, c_j, u_j and D_j are the charge, concentration, mechanical mobility and diffusion coefficient, respectively, for species j. N is the number of species and ϕ is the electro-chemical potential. Let's define the conductivity of the electrolyte as the following:

$$\kappa \equiv F^2 \sum_j z_j^2 u_j \tag{17}$$

The conductivity κ has the dimension $[\Omega^{-1}m^{-1}]$ or $[Sm^{-1}]$. Substituting the above definition to (16), the equation reduces to:

$$i = F \sum_j z_j D_j \mathrm{grad} c_j - \kappa \mathrm{grad}\phi \tag{18}$$

The first term of (18) corresponds to the current density sustained by concentration gradients. This term can be neglected in large-scale simulations, because concentration gradients exist only in the diffusion layer which is very thin compared to the size of simulation domain. These concentration effects are taken into account as polarization curves, and applied as boundary conditions. Finally, the current density in the electrolyte domain Ω is expressed by the following:

$$i = -\kappa \mathrm{grad}\phi \tag{19}$$

Assuming that there is no accumulation or loss of ions in the bulk of electrolyte, the following conservation equation should be required:

$$\mathrm{div} i = -\mathrm{div}\left(\kappa \mathrm{grad}\phi\right) = 0. \tag{20}$$

In case of the conductivity of the electrolyte, such as considering sea water, the conductivity κ is constant. Thus (20) reduces to a simple Laplace equation:

$$\kappa \nabla^2 \phi = 0. \tag{21}$$

It is noted that the potential ϕ is defined with referring to the metal and has the inverse sign of the employed usually in the corrosion science in which the potential E is defined to a reference electrode such as saturated calomel electrode (SCE) or (SHE).

The current density i on boundary of the electrolyte domain Ω is defined by the following equation:

$$i \equiv i \cdot (-n) = \kappa \frac{\partial \phi}{\partial n} \tag{22}$$

where n is outward normal vector, $\partial/\partial n$ is the outward normal derivative. It is noted that i is positive when the current flows into the electrolyte from the boundary.

Let Γ be the whole surface of the electrolyte domain Ω as shown in fig. 7. The boundary Γ consists of the following three types of boundarys:

- The Neumann-type boundary Γ_n on which the current density i is prescribed.
- The Dirichlet-type boundary Γ_d on which the potential value ϕ is prescribed.
- The metal surface Γ_m on which the relationship between the potential value phi and the current density i is prescribed with the polarization curve.

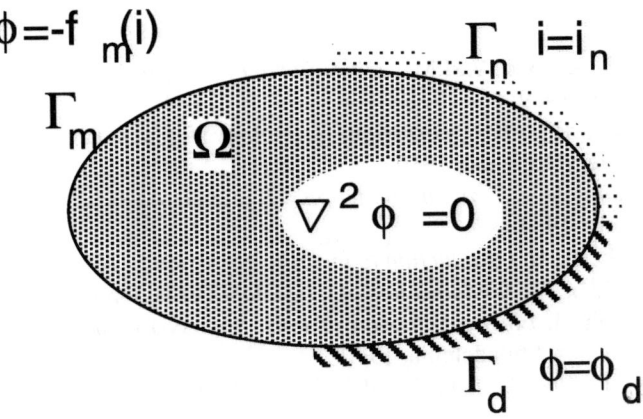

Figure 7: The governing and boundary condition.

In case of considering the surface of sea water or electrically insulated material such as plastic resin wall, the current density i is 0 for these boundaries, so that they can be modeled with the Neumann-type boundary Γ_n. The mathematical form of the above three types of boundaries can be expressed as the following:

$$i = i_0 \qquad on \; \Gamma_n \qquad (23)$$

$$\phi = \phi_0 \qquad on \; \Gamma_d \qquad (24)$$

$$-\phi(\equiv E) = f_A(i) \qquad on \; \Gamma_A \qquad (25)$$

$$-\phi(\equiv E) = f_C(i) \qquad on \; \Gamma_C \qquad (26)$$

where, i_0 and ϕ_0 are the prescribed value, respectively. $f_A(i)$ and $f_C(i)$ are the non-linear functions representing the experimentally determined polarization curves for anode and cathode, respectively. Therefore, the corrosion rate analysis reduced to solving the boundary potential problem of which (21) is the governing equation and (23)–(26) is the boundary condition.

References

[1] Fontana, M.G. & Greene, N.D., *Corrosion engineering*, McGraw-Hill: New York and London, 1978.

[2] Aoki, S. & Urai, Y., 'Inverse analysis for estimating galvanic corrosion rate', *IUTAM Symposium on Inverse Problems in Engineering Mechanics*, 1992.

[3] Amaya, K., Togashi, J. & Aoki, S., 'Inverse analysis of galvanic corrosion using fuzzy a priori information', *Japan Soc. Mech. Eng., Ser. A*, **38**, pp. 541–546, 1995.

[4] Miyasaka, M., Amaya, A., Kishimoto, K. & Aoki, S., 'Boundary Element Analysis on Corrosion Problems of Pump and Pipe' *Proc. the 1995*

ASME/JSME Pres. Vessel and Piping Conf., ASME PVP, **306**, pp. 279–284 1995.

[5]   Mills, R. & Lobo, V.M.M., *Self-Diffusion in Electrolyte Solutions*, Elsevier: New York, 1989.

[6]   Mattsson, E., *Basic corrosion technology for scientists and engineers*, Chichester: Horwood, 1989.

[7]   *Corrosion basis: an introduction*, National Association of Corrosion Engineers, 1984.

[8]   Evans, U.R., *An introduction to metallic corrosion* E. Arnold, London: 1981.

[9]   Scully, J.C., *The fundamentals of corrosion*, Pergamon Press, Oxford: 1990.

[10] Rubinstein, I, *Electro-Diffusion of Ions*, Society for Industrial and Applied Mathematics, Philadelphia, 1990.

[11] Crank, J., *The Mathematics of Diffusion*, Oxford University Press: Oxford, 1975.

[12] Deconinck J., Current Distributions and Electrode Shape Changes in Electrochemical Systems, *Lecture Notes in Engineering*, **75**, Springer-Verlag: Berlin, 1992.

Shipboard impressed current cathodic protection system (ICCP) analysis

V. G. DeGiorgi[1], E. Hogan[2], K. E. Lucas[3] & S. A. Wimmer[1]
[1]Multifunctional Materials Branch, Naval Research Laboratory, Washington, DC.
[2]Material Science and Technology Division, Naval Research Laboratory, Washington, DC.
[3]Center for Corrosion Science and Engineering, Naval Research Laboratory, Washington, DC.

1 Introduction

In technical work, the world is often described by one of three sources of values; real structure measured response, experimental measurements or calculated computational results. While it is common to discuss results from each of these three methodologies as representing the same condition this is not really the case. Actual measured structural response is the global behavior of a component based on the complex interaction of geometry, materials, boundary conditions, environmental factors and applied loads. Models, either computational or experimental, are simplified representations of the actual structure. In many instances the accuracy and acceptance of experimental and computational methods is based on how well these methods can duplicate the measured response from a real structure. While it is recognized that both experimental and computational modeling must be initially validated by comparison with rather simplistic structural systems, the emphasis must be on simplified structures. If more complex structures or systems are used as validation problems then issues such as real versus conceptual definitions of boundary conditions, loads and material response must be addressed. These issues are often at the core of the complex nature of structural response. Once an experimental or computational methodology has been validated through a series of comparisons with controlled simple structure or system performance predictions, these methods should be considered as equally 'true' in their results as measured data from a real structure is considered 'true'. Important in evaluating any measured or calculated data is an understanding of the structure or system and associated loads, boundary conditions, and material response definitions.

Rather than consider measurements from real structures as 'truth' values and data from modeling as a second best information source, the authors are proposing that each source provides essential insight into how an actual system will perform under a variety of conditions. Insight can be obtained on the underlying physical phenomenon from the more controlled situations of experimental and computational models. Experimental models can be used to help identify issues and further the understanding of computational models. Computational models can be used in the same way to further the understanding of an appropriate application of experimental models. Results from both are combined to better understand the complex real world response.

Modeling is more than just a representation of the actual structure. It is a methodology for advancing the understanding of system performance. It is a methodology for determining relative importance of the many factors that have an influence on system performance. Each type of modeling can be presented as a different means to represent reality. What is crucial in this approach is not only having a technical and theoretical grasp of the modeling technique applied but to have knowledge of and understanding of the real world system being modeled and the complimentary modeling technique. Analysts performing computational modeling cannot be effective without an understanding of the experimentalist's world. The same understanding of the computational approach is required of the experimentalist. The authors are not proposing that each be a specialist in the other modeling methodology. Rather a general working knowledge and understanding of basic approaches and limitations is required. Both analyst and experimentalist also need a working knowledge of the system they are modeling. Simplifying assumptions are standard practice in the creation of computational and experimental models. The better the understanding of the actual system, the better the choices that can be made on the necessary simplifying assumptions.

The authors realize that requiring this cross over of knowledge is not standard operating procedures in many facilities. However, it is an approach that has worked successfully in the past and it is an approach that allows for true multidisciplinary research. Problems in today's world are complex. There is seldom a simple answer. Conflicting requirements due to system performance requirements, material limitations and other issues have driven design problems to the realm of multidisciplinary research. Unified or multidisciplinary design approaches provide for the most effective solution.

The authors have applied this approach to the design and evaluation of shipboard impressed current cathodic protection systems. In a multidisciplinary unified approach an understanding of real structure, experimental methodology and computational methodology is required. This information is presented. This chapter provides detailed information on the development of the computational modeling methodology applied to these systems. In addition, real system information and an introduction to the physics basis for experimental methodologies are provided. Chapter sections are:

Cathodic protection systems – a basic introduction
Ship systems
Experimental modeling
Computational modeling – governing equations
Computational modeling – ICCP systems
The road ahead – unified design approach

The process and information provided is focused on modeling of cathodic protection systems, however, the approach can be applied to any system or structure. It is critical to understand that experimental, computational and real structure data all provide information required to further the understanding of the system performance in changing conditions.

2   Cathodic protection systems – a basic introduction

In a floating structure, the external wetted hull is generally considered to be the cathode or the working electrode, which requires protection. Anodes are co-located on the underwater hull and output current, from the power supplies, which effectively flows through the electrolyte, to the hull, to complete the electrical circuit. Current flow within the electrolyte is accomplished through the utilization of electrons in the electrochemical reactions at the surfaces of the anodes and corresponding cathodes. Where current is impressed on the cathode, the cathode potential is driven in a negative direction which can be measured using standard Silver-Silver Chloride (Ag/AgCl) reference cells. Hull potential measurements provide information concerning relative current distribution from the anodes and a determination of the effective levels of polarization to different areas.

A shipboard ICCP system has three basic components: 1) reference cells, typically Ag/AgCl, 2) controller/power supply, and 3) anodes. Typically marine ICCP systems utilize reference cells to monitor potential levels at critical locations on the hull and to provide an electrical feedback to the controller circuitry which regulates the output current to the anodes. In this way, the ICCP has self-regulating set potential levels for operation, normally between –0.80V to –0.85V versus the Ag/AgCl reference cell, which is sufficient to protect steel structures. The ICCP system also naturally compensates for most environmental factors, which influence the cathodic current demand behavior, by continually regulating the anode current output to maintain the hull polarization to the designated set potential.

The ICCP system in principle works well for this task, but the key to a good design and well functioning system is the proper location of components on the complex hull geometry at the onset. Until 1986, ICCP systems were designed empirically and typically were unable to respond to the variable life-cycle conditions, which resulted in hulls that were either overpolarized, underpolarized or both. Either extreme is dangerous for a hull, because overpolarization tends to cause excess gas evolution and alkali damage that cathodically disbonds the barrier coatings, while underpolarization allows the materials to freely corrode and/or galvanically interact. The complex hull geometry requires that anodes be located such that the current distribution can be uniformly maintained throughout and correspondingly that the reference cells are placed such that they effectively monitor the hull potential at necessary points.

A schematic of a typical analog ICCP system for a larger ship hull is shown in fig. 1. Often an ICCP system is divided into zones, which consist of a group of anodes, controlling reference cell and associated controller/power supplies. This zone operates essentially independently from other zones, except that the hull is the common ground point and thus each system can influence the operation of the other, depending on system design. The advantage of a zone system, however, is that the reference cells can be placed more proximate to those areas which require protection and the protection can be focused towards more localized hull areas, such as the stern area, which has complex hull, rudders, struts and propellers. More recently, advanced digital ICCP systems have been developed, which are software driven and are less zone oriented. A schematic example of a software controlled system is shown in fig. 2. These systems have the same basic components, and can be setup exactly like a zone system or anode output can be based on a composite of reference cell readings with an algorithm determining individual anode current output. What is imperative to note, however, is that a poorly designed ICCP system, with poor component placement cannot be improved significantly by more sophisticated electronics.

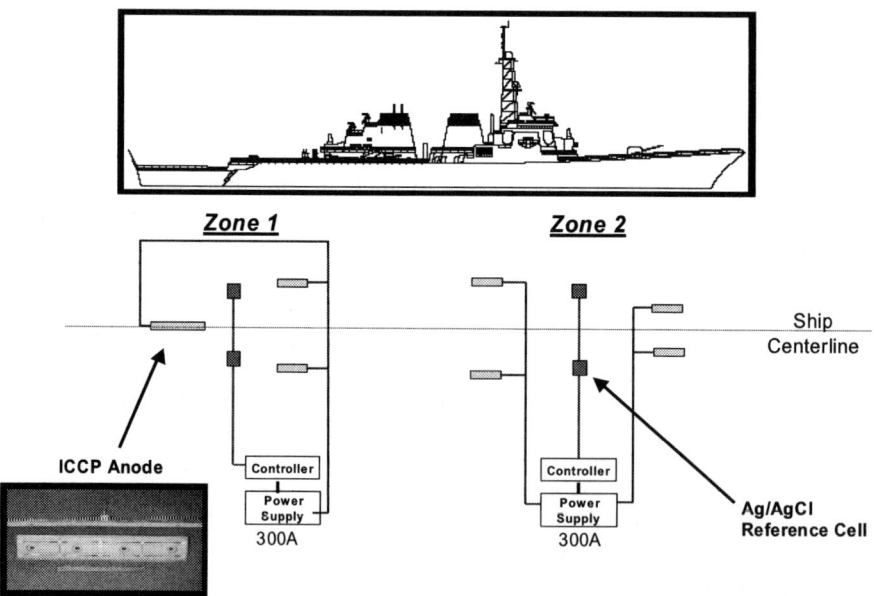

Figure 1: Typical two-zone ICCP system.

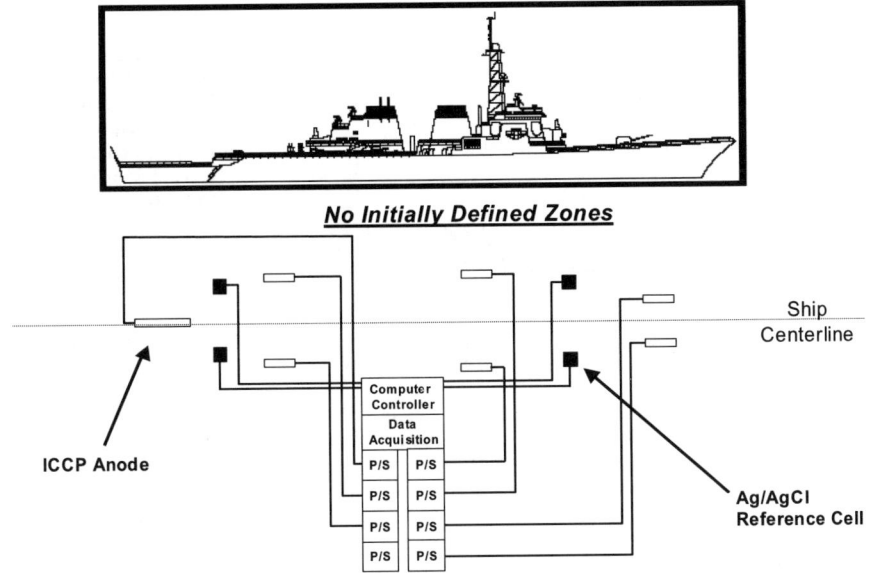

Figure 2: Schematic of the 'next generation' computer controlled ICCP system.

3   Ship systems

We often forget how complex the real world is and a ship ICCP system is no exception. Each ship has its own dynamic sequence of time, from when the hull is new to when it is removed from the water for refurbishment of the coatings. Within a typical maintenance life-cycle, each ship hull undergoes significant changes and differences within operational behavior, docking periods, coatings degradation, and biological fouling. What is acceptable ICCP system per-formance when the hull is new must also be acceptable throughout the full life-cycle experience.

Simple schematic representations of the structural component or system are often presented as representative of what is being modelled. While these may accurately represent the computational or experimental model, they are simplifications of the real world. In reality the factors which affect the corrosion behavior and signature of the underwater portion of a ship hull are numerous, often highly complex and variable. A partial list of the factors that influence ICCP system performance is shown in fig. 3.

Often measured data from a real structure is presented as a 'truth' value. This requires a detailed knowledge of the condition of the structure when the data measurements are obtained. Unfortunately a real ship, or any complex structure, can provide only limited information about its physical condition and protection requirements. Beyond the changes in performance caused by physical properties, uncertainties in boundary conditions and applied loads also have to be taken into consideration when evaluating data. Natural e nvironmental effects on materials

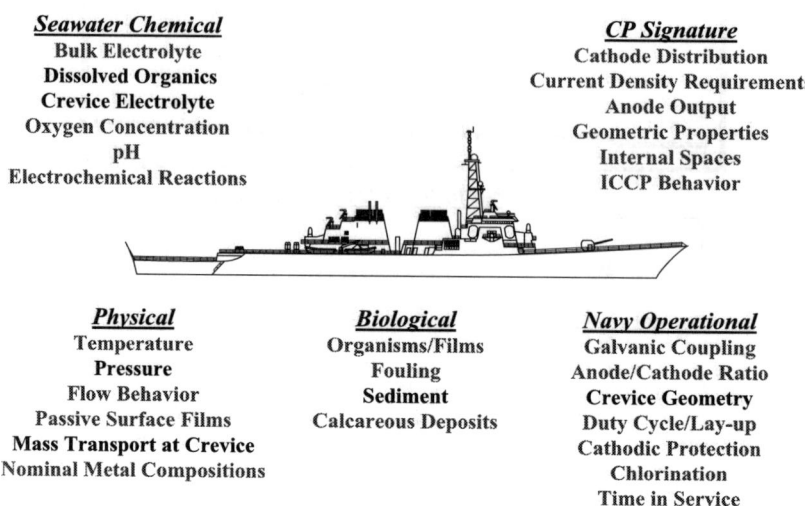

Seawater Chemical
Bulk Electrolyte
Dissolved Organics
Crevice Electrolyte
Oxygen Concentration
pH
Electrochemical Reactions

CP Signature
Cathode Distribution
Current Density Requirements
Anode Output
Geometric Properties
Internal Spaces
ICCP Behavior

Physical
Temperature
Pressure
Flow Behavior
Passive Surface Films
Mass Transport at Crevice
Nominal Metal Compositions

Biological
Organisms/Films
Fouling
Sediment
Calcareous Deposits

Navy Operational
Galvanic Coupling
Anode/Cathode Ratio
Crevice Geometry
Duty Cycle/Lay-up
Cathodic Protection
Chlorination
Time in Service

Figure 3: Factors affecting ship corrosion and signature.

also have impact on the ICCP parameters and must be taken into consideration. For example, in the case of ship hulls with a good dielectric barrier coating, it may seem appropriate to assume a total blockage of current but this would not be realistic. Over the life-cycle, these materials may adsorb water in an undefined manner, fail completely and also suffer from unpredictable changes related to mechanical breakdown and other application discrepancies. It is very seldom that a structure can be completely described.

In principle, the major factors that must be considered in protecting a hull from corrosion are: 1) the nominal wetted surface area which requires protection, 2) the material characteristics of metallic components exposed to the seawater and 3) the chemical aspects of the bulk electrolyte (seawater) under operational conditions, such as, seawater conductivity, pH, dissolved oxygen and surface reactions. Directly influencing the kinetic behavior of these reactions are the temperature, velocity and diffusion properties of the surfaces, both with and without cathodic protection applied. In addition to the growth of calcareous deposits while cathodically protected, all surfaces may foul with marine organisms, resulting in a biological system that will further influence the surface properties of the metals involved.

In many cases, unique operating conditions of the ship will have a direct interplay on the ICCP system performance, because at some point the ship may simply turn the system off for an unknown period of time or move rapidly into greatly different waters or nest alongside piers, ships or other geometric features. When nested or in fresh water the ICCP system performance can vary significantly. Hull geometry and composition may further impact the performance by changing flow regimes, eroding surfaces and by creating galvanic corrosion problems, especially at the propeller areas.

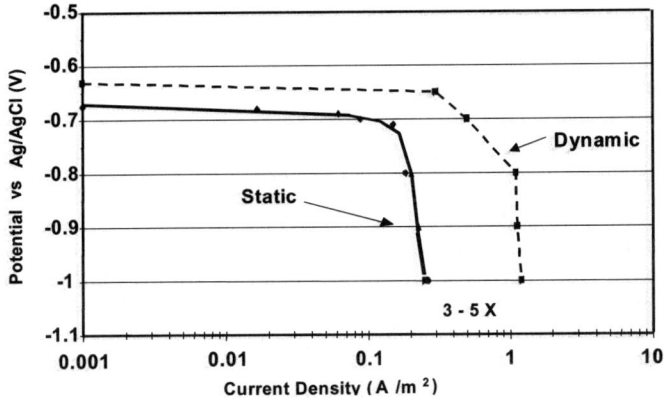

Figure 4:   Steel current density requirements as function of seawater flow.

Material interaction is a complexity that exists for most shipboard ICCP systems. In reality there are multiple exposed materials on a ship, not a single material, *i.e.* steel, which require protection. The multiple material systems result in numerous galvanic corrosion problems, most notably the galvanic couple between the steel hull and the nickel aluminum bronze propellers. These galvanic material relationships, while protected under ICCP, greatly influence the current density requirements, location of components and system operation. Structural characteristics also add to the complexity by hosting other types of corrosion, such as crevice or hydrogen effects.

In determining what simplifying assumptions can be tolerated, one must consider the dominant factors in performance. For corrosion protection, bare metallic areas become the dominant corrosion protection issue on any hull and it is these cathodic surfaces (bare areas), rather than any anodic surface that define the behavior of the ICCP system. This is because the system power supplies can easily overwhelm any anodic resistance and will polarize the cathodic surfaces in response to the reference cell feedback to the controller. The rates of corrosion at the cathode are influenced most by the ability to diffuse oxygen across the metal/electrolyte interfaces. This effect is shown in fig. 4 and it can be seen that the relative differences between a static current density requirement and a dynamic state may increase the demand by a factor of three to five times. While the ship is in motion, oxygen transport is significantly increased, but the surface diffusion effect may be decreased depending on the extent of corrosion deposits, prior calcareous deposition and macro-fouling accumulation.

As stated at the beginning of this section, we often forget the level of complexity in the real world. Obviously everything cannot be captured by experimental procedures or included in a computational model. The challenge for modelling, both experimental and computational, is to provide an accurate representation of the many localized environments on the ship hull and to accurately represent the many different operational environments that the ship will experience. Judicial use of simplifying assumptions is required. The better the entire problem is understood, the closer the final modelling environment created for analysis will be to the real world.

4   Experimental modeling

Physical Scale Modeling (PSM) is the name applied to a technique that has been developed that uses near-exact scale models and scaling factors based on the physics of electrochemical response to provide information on current and potential values on the structure. The near-exact scale models, ranging in length from two to ten feet, such as shown in fig. 5, are the representative geometry that are used along with scaled electrolyte to provide equivalent ohmic paths. Basic aspects of potential and current distributions in scaled systems were originally presented for use in electroplating but can be applied to cathodic protection systems. Works by Kasper [1], Agar and Hoar [2] and Weber [3] define the relationship between scaled geometry, scaled solution conductivity and the resulting interpretation of results. These works can be directly related to the modern practices of PSM as practiced by NRL Center for Corrosion Science and Engineering. Further theoretical basis for PSM has been presented by Ditchfield *et al* [4]. Validation of the application of PSM to surface ship by means of comparison with real ship data has been documented by [5, 6, 7].

PSM was pursued specifically to provide a robust physics based method for the design of ICCP systems. A driving force behind much of this work is the U.S. Navy requirement of ICCP as the standard method to achieve good corrosion protection of the hull. The PSM technique has been utilized to determine the best ICCP component placement, life-cycle performance, zone interaction behavior and various failure modes under both static (dockside) and dynamic (underway) operational conditions for a variety of ship and system configurations. PSM can directly address difficult geometries, areas of restricted flow, protective coatings degradation, advances in ICCP design technology

Figure 5:   Example of a near-exact scale model used in PSM.

(*i.e.* use of digital comptrollers – hardware and software, advanced control algorithms) and complex interaction of power supplies and control algorithms (*i.e.* zone behavior). The U.S. Navy has utilized PSM as performed by NRL Center for Corrosion Science and Engineering extensively. There is an established design criteria and PSM is currently the accepted standard technique for design of U.S. Navy ICCP systems [8].

For PSM and the computational techniques, cathodic protection behaves in accordance with Ohm's Law:

$$E = I (R_P + R_{OHMIC})$$ (1)

where, E = potential (V), I = current (A), R_P = polarization resistance and R_{OHMIC} = electrolyte ohmic resistance. In scaling, necessary in the PSM technique, $R_{OHMIC} = \rho L/A$, and where, ρ = electrolyte resistivity, L = length of ship or model and A = area. For exact scaling, it is desired for potential relationships to exist, such that $E_{SHIP} = E_{MODEL}$ and for current density (i) behavior, such that $i_{SHIP} = i_{MODEL}$, where i = A/m^2. For a relationship where:

$$E_{SHIP} = I_S(\rho_S L_S/A_S) = i_S (\rho_S L_S)$$
$$E_{MODEL} = I_M(\rho_M L_M/A_M) = i_M (\rho_M L_M)$$ (2)

It is necessary that $R_{P\ (SHIP)} = R_{P\ (MODEL)}$ for the model to scale exactly, by definition. For scaled models $L_S/L_M = k$ and $\rho_M = \rho_S(k)$, where k = scale factor, the relationship becomes:

$$E_{SHIP} = E_{MODEL} = i_S (\rho_S L_M)(k) = i_M (\rho_S L_M)(k)$$ (3)
$$i_S = i_M$$

For PSM current measurement on the model, it follows from eqn (2) and eqn (3) that:

$$I_S = I_M(k)^2$$ (4)

Basic assumptions in the derivation of the modeling process provide for precision measurement of potential and current on the model with a direct mathematical relationship between the model and full scale system. These assumptions are:

• The surface areas and geometry are exact and scaled such that $A_{SHIP} = A_{MODEL} (k)^2$.
• The current density relationship, $i_{SHIP} = i_{MODEL}$ is true, by definition, when the model size, electrolyte dilution and polarization resistance components obey the scaling law.
• $R_P = \Delta E/i_C$ must be same for the model and full scale system, where ΔE represents the polarization from E_{CORR} to the cathodic protection set potential of $-0.85V$.

The key to the modeling methodology, in addition to the correct implementation of model size and correct electrolyte ohmic scaling is the polarization resistance behavior. In order to preserve the relationship $R_{P\ (SHIP)} = R_{P\ (MODEL)}$ for direct scaling, NRL has established a pre-conditioning sequence. In pre-conditioning the models are first cathodically protected in full scale seawater (to $-0.85V$ Ag/Ag Cl electrode), to allow the deposition and adsorption of natural calcareous films on the model surfaces prior to testing. It has been experimentally shown that once a calcareous film was deposited, prior to placement in the scaled electrolyte, the R_p component of the metallic surface would behave in a manner very similar to natural seawater behavior and that the scaling equation was correct. Measurements on the model produced correct potential values, correctly scaled current behavior and normal cathodic surface responses. When operational conditions in the experimental facility are varied, such as velocity, coatings damage or anode/cathode ratios, the system compensates in a natural manner because it is in dynamic equilibrium. For computational modeling measurements, this critical parameter is controlled by the use of polarization curves for each of the ship states, materials properties and hull conditions.

Like the real ship, the PSM technique utilizes a controlling reference to provide a potential set point and feedback circuit for control. Accordingly, the system current output can be monitored, but instead of dealing with an unknown hull surface state, the model is well defined into specific cathode areas which have independently monitored/controlled cathodic behavior, monitored anode sites and multiple additional hull potential sensing points from the reference cell array. Thus, in addition to what would be basis ship data, the model provides detailed cathodic current demand, anode currents outputs and an array of potential data, for a variety of coatings damage conditions, under static and dynamic states. Outside of the rules for scaling defined above, assumptions in the PSM technique for ICCP design lie primarily in the selection of the metal percentages for the cathode areas and in the selection of the design current densities. Electric field (EF) behavior produced as a result of the ICCP current flow through the test electrolyte can also be rigidly evaluated using reference cell pairs located discretely or scanned under the model. An example of the electric field signature obtained during the PSM process is shown in fig. 6. Results for EF calibration, as compared to theory for dipole measurements, are exceedingly accurate and hull signature can thus be evaluated in real time.

All methodologies have limitations. The limitations associated with the PSM technique tend to be associated with the mechanical aspects related to PSM modeling. For instance there are physical limitations in defining a detailed model and reduction in current density demand due to scaling factors. Financial and time limitations are also predominately associated with the mechanical aspects of PSM. Limiting factors are the ability (cost and time) of procuring a detailed model, the time associated with the natural polarization changes and setting up the model to perform the iterative tasks associated with optimizing component placements. Experimental control in the modeling is maintained by testing only when the cathodic current density requirements are within the defined protocol ranges. This results in testing conditions that are always repeatable, but it also results in the model falling out of the required test range and therefore requires

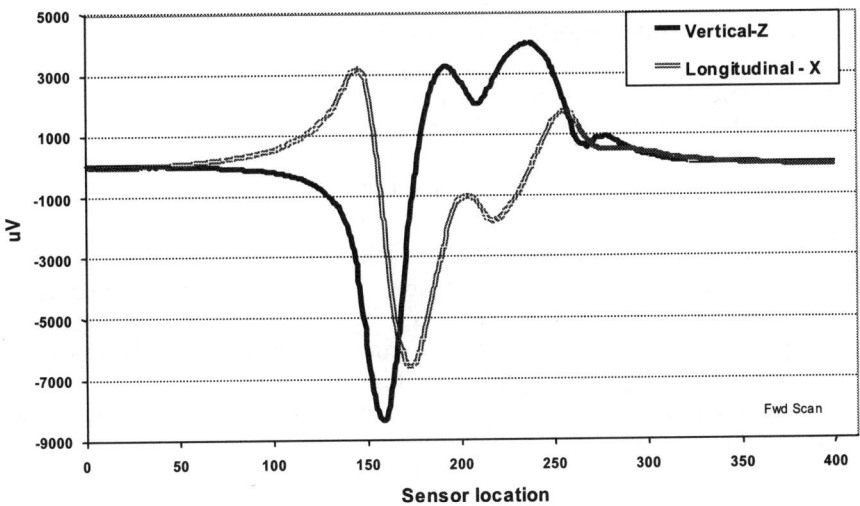

Figure 6:   An electric field scan obtained during the PSM process.

periodic re-conditioning of the surfaces. The PSM process is schematically represented in fig. 7. For many of the cumbersome and time-consuming tasks, the computational modeling technique offers an immediate enhancement from the iterative design repetition of PSM and can be a highly beneficial and cost effective partner in a comprehensive design and verification process. This is one way in which the two modeling techniques result in complimentary information gathering with the end result being more detailed information to help in determining system performance characteristics.

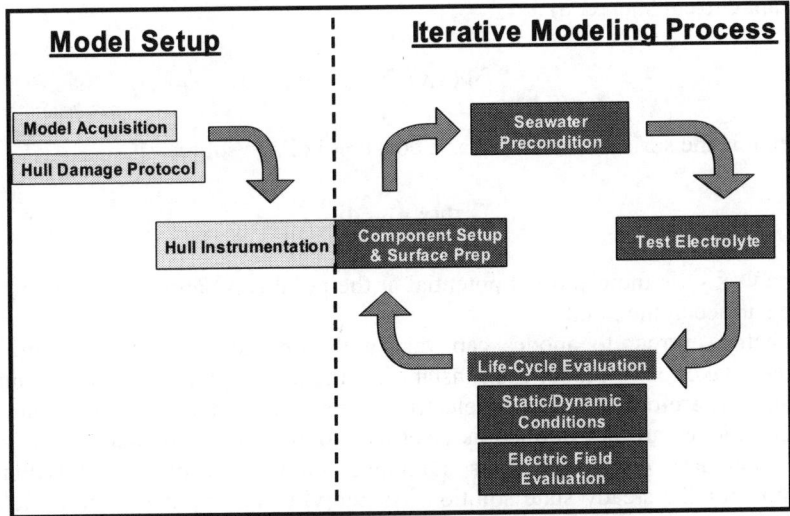

Figure 7:   PSM technique iterative process.

5   Computational modeling – governing equations

The governing differential equation for electrochemical corrosion for a structure surrounded by a bounded uniform electrolyte is:

$$k\nabla^2\Phi = 0 \tag{5}$$

Where Φ is the electric potential and k is the conductivity of the electrolyte in domain Ω. For Laplace's equation to be valid the volume surrounding the structure cannot contain either electrical sources or sinks and the total current in must equal the total current out of the system. The equation models steady state conditions and does not address corrosion initiation.

The solution space for the problem is the surface Γ, which bounds the domain Ω and is defined as:

$$\Gamma = \Gamma_A + \Gamma_C + \Gamma_I \tag{6}$$

Where A, C and I are anodic, cathodic and insulated regions. In the case of a shipboard ICCP system the surface is the wetted surface of the ship's hull. The domain is the open sea surrounding the ship. It is assumed that the seabed is a large distance removed from the water surface and lowest point of the ship's hull. Interactions between the hull and other objects, such as a second ship or pier, or between the hull and seabed are not addressed by the presented analysis method.

The surface is divided into anodic regions, cathodic regions and insulated regions. Each region represents a specific component of the ICCP system and ship hull. Anodic regions are the ICCP system anodes and are defined as either a constant current source, q_A:

$$\partial\Phi(x,y)/\partial n_{(x,y)} = q_A \tag{7}$$

Where n is the surface normal, or as a constant voltage source, Φ_A:

$$\Phi(x,y) = \Phi_A \tag{8}$$

Where $\Phi(x,y)$ is the electrical potential at the point (x,y) and $n_{(x,y)}$ is the normal to the surface at the point.

Electric current to anodes can vary with time for an ICCP system. The approach taken is to model each instant in time as a separate boundary element solution. Therefore changes in electric current are not incorporated into the model. Anodes are not defined as electrical sources in the model but as fixed-value boundary conditions. This maintains the validity of use of LaPlace's equation for the steady state solution. An individual computer solution defines steady state conditions at a specific point in time at a specific anode current level.

The electric flux at a point on the surface of the cathodic region is defined by:

$$\partial\Phi(x,y)/\partial n_{(x,y)} = f_c \tag{9}$$

Where f_c is the cathodic polarization function. The polarization function is experimentally determined and is typically non-linear. In the current analysis, polarization response is modeled as piece-wise linear through use of a look-up table format. In general, if the value of the potential is defined the current can be determined and if the value of current is defined the value of potential can be determined.

At insulated surfaces, such as typically used to define painted surfaces, the flux is constant through time and equal to zero:

$$\partial\Phi(x,y)/\partial n_{(x,y)} = 0 \tag{10}$$

A 'zone' is a power supply. The anodes attached to that power supply and the associated reference electrode(s) comprise a zone. Each zone has its own control algorithm. In the computational work done to date the control algorithms have been simple feedback control based on the potential at the reference electrode(s). Each power supply is controlled independently. Anodes are defined by assigning specific elements in the boundary element model potential, current or current density values. Mathematically the solution does not matter on the choice of boundary conditions to define the source anodes. In all cases anode values are defined as input values. Reference electrode values are determined in the solution process and are not boundary condition defined values.

Input values, boundary conditions and material properties are combined for the mathematical solution of LaPlace's equation as represented by eqns (5)–(10). In addition to mathematical criteria for a solution there are two additional criteria for a feasible solution:

(1) Potential values at the nodes representing the reference electrodes equal to the target value;
(2) Total current to anodes associated with a single power supply is less than the power rating of that power supply.

A candidate solution consists of a computer run in which the potential of the reference cells is at the target potential –0.85V Ag/AgCl electrode (criteria (1)). Reference cell readings are calculated potential values. A feasible solution occurs when the reference cell is at the target potential (criteria (1)) and the total power required is within the power supply capacity as defined in the ICCP system design (criteria (2)). This is not a constraint required by LaPlace's equation or the representations used for material response. A feasible solution is determined through a multiple run process in which anode input values are varied. A feasible solution is typically the solution of interest.

There are two basic material characterizations required for evaluation of ICCP systems: the conductivity of seawater and the polarization response for all materials used to define the hull. Seawater is treated as a uniformly mixed solution with a single value of conductivity for all depths. The accuracy of the polarization data used to define material response is critical to the overall solution accuracy. In a perfect world the modeler would be able to define material properties that were determined from conditions that exactly duplicated the environment of interest. This unfortunately is not generally the case. Most polarization data are obtained from small specimens, tested in the laboratory, without the influence of material interactions and under controlled flow conditions. As in all laboratory testing, the tests define material response in a very controlled situation. Even with good control it has been determined that material polarization response is highly sensitive to many parameters. Engineering judgment is used to determine the best fit between service and material characterization test conditions.

There are four distinct material definitions that have been used in the NRL computational analysis work: perfect paint, finite resistance paint, steel, a nickel aluminum bronze alloy (NAB) and seawater. Perfect paint is by definition a perfect insulating surface. Perfect paint is not defined using a polarization curve as input but is defined using boundary conditions. The other three materials are defined using nonlinear polarization curves as input data. Polarization responses for steel and NAB can be found in the literature. Values used in the NRL analyses were those experimentally measured specifically for projects related to PSM. The performance of a 'real' paint has been modeled by shifting the polarization response of steel by three orders of magnitude on the current demand axis. Paint response is discussed in more detail later in the chapter.

Despite the best estimates of material response, boundary conditions, ICCP system anode performance and geometry definitions, it is essential to remember that these methods are modeling techniques that require simplification from the real world. This is not done in ignorance or with any intention to misrepresent the structure or system and its environment. In fact, defining what factors are simplified is often the most challenging aspect of setting up a new computational model. The goal is to create the most accurate representation of the real world model while remaining within computational system or operating defined limitations. One often hears those involved with computational modeling discussing element type and number limitations. Even with today's high-speed computers, large storage memory capabilities and advances in solution algorithm complexity, the complete real world picture cannot be incorporated into a model.

Another factor that is important for the analyst to remember is that, typically, operating parameters in the real world are defined in ranges rather than point values. Maximum and minimum calculations based on key conditions are necessary to create a range of operation performance. In computational models where the end goal is to determine maximum structural performance the concept of range can be collapsed by collocating worse case conditions (*i.e.* maximum possible load, minimum dimensions and minimum material strength) in the computational evaluation. The resulting calculations can be considered upper bounds in terms of deflections, strains and stresses. If these values are within safe margins then all anticipated operating conditions will be within safe margins of operation. In general, in the design of ICCP systems, this collapsing of ranges into a single point is not the desired approach. The limiting value that must be applied to all conditions is the power supply capacity. This value, when calculated, should be done using worse case combinations. Otherwise, the power required and voltage set points for protection at one operating condition are of equal interest as those at another slightly different operating condition. Beginning of life and end of life damage conditions are used to bound system performance. Ultimately the end-user is interested in a range of operating parameters corresponding to uniquely identifiable operating conditions.

6   Computational modeling – ICCP systems

The mathematical development of boundary element techniques and aspects of the method as applied to electrochemical corrosion systems have been traced by other researchers. ICCP systems are based on the electrochemical corrosion phenomenon. Review articles by Adey [9], Munn [10] and Gartland *et al* [11] have clearly shown the case for the usefulness of the boundary element method. Some computational codes, such as described in [12], are based on an analogy between electrochemical corrosion and heat conduction. Other commercial programs and specialized codes directly solve LaPlace's equation for steady state corrosion.

The boundary element technique has been applied to a variety of cathodic protection systems. Structures studied range from pipelines that can be represented by 1D models to geometrically complex structures such as ships that require full 3D modeling. Pipeline analyses that examine the electrochemical corrosion behavior as well as determining the effectiveness of cathodic protection systems have been presented from the mid-1980s on. Bardal *et al* [13] examined the behavior of carbon steel and stainless steel components. Sacrificial anode configurations for pipelines have been studied by Adey [14]. Brichau and Deconinck [15] evaluated parallel pipes buried in the ground. Lee *et al* [16] evaluated the effects of a non-uniform electrolyte composition on the protection system of a buried pipeline. Yan *et al* [17] and DeGiorgi [18] discussed galvanic material couples in straight piping systems. Amoya and Aoki [19] used boundary element techniques to determine the optimum anode locations in storage tanks. The use of boundary element techniques to model offshore oil structures is well established [20]. Offshore structure modeling is unique in that polarization response can be determined from *in situ* measurements eliminating issues related to laboratory development of polarization data. Significant work, as collected in [21] has been completed on ship systems. Work continues to date by many researchers on ICCP systems. Diaz and Adey [22] have recently presented a methodology based on boundary element techniques to determine the optimum anode configuration for shipboard ICCP systems.

Since 1987 NRL has actively pursued computational modeling of shipboard ICCP systems using boundary element techniques. Issues addressed include initial proof of concept modeling and validation of the approach through comparison with PSM experimental results. The work performed, once the process was validated, branched into evaluating basic modeling boundary condition assumptions as well as geometric simplifications. Later work has addressed issues related to PSM experimental work. The commercial boundary element codes BEASY-CP [23] and Frazer-Nash Detailed Modeler [12] have been used for the analyses presented. The commercial code PATRAN [24] has been used for model generation. The commercial code TECPLOT [25] has been used for results visualization. In addition, customized computer programs for translation and display of data have been developed at NRL. Modeling results and guidelines generated from the analyses are applicable to any code used. BEASY-CP solves LaPlace's equation for steady state corrosion. The Frazer-Nash Detailed Modeler uses the thermal-electrochemical corrosion analogy. Guidelines have been found to be equally valid for both codes.

The boundary element problem requires a mathematical representation of the outer surface of the ship hull and the surrounding volume of electrolyte. The ship hull is modeled by a boundary element mesh. The surrounding volume of electrolyte can be modeled in one of two ways; either by a mathematical boundary condition that defines an infinite or semi-infinite volume or by an outer meshed surface that defines a volume. NRL has typically opted to use a large but finite volume of seawater to represent the open ocean around the ship hull. This domain is defined sufficiently large enough so that edge effects on the potential profile of the surface ship are negligible. Typical domain dimensions are 12 to 20 ship lengths. All analyses reported to date have been interested in the open ocean environment.

In all NRL studies symmetry conditions have been invoked for both the ship hull and to represent the water surface. Ship hulls and ICCP systems were defined as symmetric with respect to port and starboard characteristics. This was done in the interest of saving computational time and resources. There is no requirement for symmetry. Symmetry conditions were also used to define the water surface. This is a standard approach in boundary element methods.

A basic design matrix consisting of four cases was created by the pairing of two service flow conditions, static and dynamic, with two paint damage conditions. Static flow represents dockside conditions. Dynamic flow represents ship underway conditions. The two paint damage conditions are minimum (2.8% of the hull surface area is damaged paint) and maximum (15% of the hull surface area is damaged paint). The location and size of damaged paint regions is defined by protocols provided by NRL Center for Corrosion Science and Engineering and Naval Sea Systems Command. Damaged paint areas are defined as exposed metal surfaces in the boundary element models. This duplicates the conditions in PSM where painted surface is represented by fiberglass and damaged paint areas are represented by strips of uncoated metal attached to the model hull.

Reference cells and anode locations in the computational model duplicate as close as possible the locations in the PSM models. In cases where port and starboard anode locations are not symmetric, the boundary element model anode is placed at the average of the port and starboard locations. The decision to use port-starboard symmetry in the computational model was based on model size and existing computational resources when the analyses were initiated.

The boundary element mesh of the ship hull is the geometric representation of the actual ship structure. Levels of detail and accuracy in geometric representation, as well as appropriate use of boundary conditions and material definitions, will have a direct influence on resulting computational accuracy. A series of guidelines for analyses have been established based on the body of work performed. These guidelines, divided into 3 categories, are:

- Model Definition.
  - A more refined model is needed than is traditionally associated with boundary element techniques.
  - Accurate modeling of relatively small-scale features, such as bilge keels, is necessary.
  - Propellers can be represented by thin disks. Further detailed modeling of this feature is a goal of future work.
  - Variations in seawater conductivity that correspond to changes in deployment region can be significant to system performance and should be incorporated into the design basis.
- Material Definitions.
  - The accuracy of computational results is directly dependent on the accuracy and appropriateness of the polarization data used as material characterization input data.
  - Preliminary design and trend studies can be successfully completed using less than optimum polarization data. Trends in performance can be determined even though magnitudes will be suspect.
  - Modeling damaged paint as totally bare metal is a conservative approach.
- Boundary Condition Definitions.
  - Modeling paint as a perfect insulating material is acceptable depending on the accuracy of results required.
  - Experimental tank size influence on results can be evaluated prior to testing.
- Determination of scaling factor prior to experimental work can be based on tank sizing considerations.

Analysis results that support each of these guidelines will be presented in the next sections.

6.1 Model definition

Initial analyses performed by NRL addressed whether boundary element techniques could be used to accurately predict system performance [26, 27, 28]. The hull geometry investigated was a U.S. Navy CG hull class destroyer. Three different ICCP systems were evaluated. Mesh refinement as well as the level of detail required were identified early as critical modeling issues. The ability to use boundary element techniques for both detailed design and trend analyses, relying on different levels of mesh detailing and refinement, was also identified as capability.

Table 1: Current demand (Amps) for CG analysis. Three measurements for evaluating results; total current to components (props. and docking blocks), current from forward and aft systems and total current. Reference cell reading = −0.85V Ag/AgCl

	Props	Docking Blocks	Forward System	Aft System	TOTAL
Min.Damage Calculated	50.3	13.7	22.3	41.8	64.0
PS Model	44.5	14.1	25.8	39.1	64.9

The original model used in the CG analysis consisted of 573 rectangular elements and yielded unsatisfactory results. Initial results were encouraging but it was felt that closer agreement between experimental and calculated results could be achieved. Two issues were identified; mesh refinement and polarization response input data. A mesh refinement study demonstrated that a significantly higher degree of mesh refinement was required. The mesh refinement study was determined to be complete when a level of refinement was determined that did not produce any additional changes in calculated results with additional refinement. Calculated results were compared with calculated results.

The original model only included propeller and rudder representations. In reviewing calculated potential profiles it was determined that there was a need to include the bilge keel. This was consistent with early model development results for PSM. After the level of mesh refinement was determined, a 3D representation

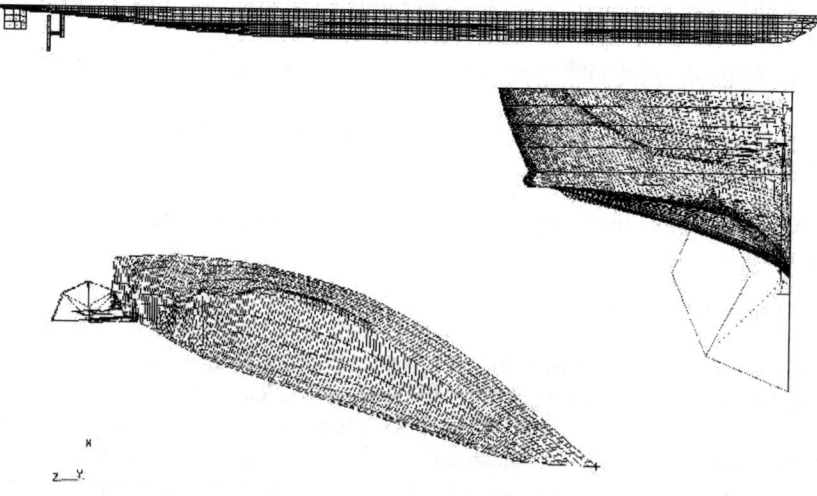

Figure 8: Boundary element mesh for CG hull class ship.

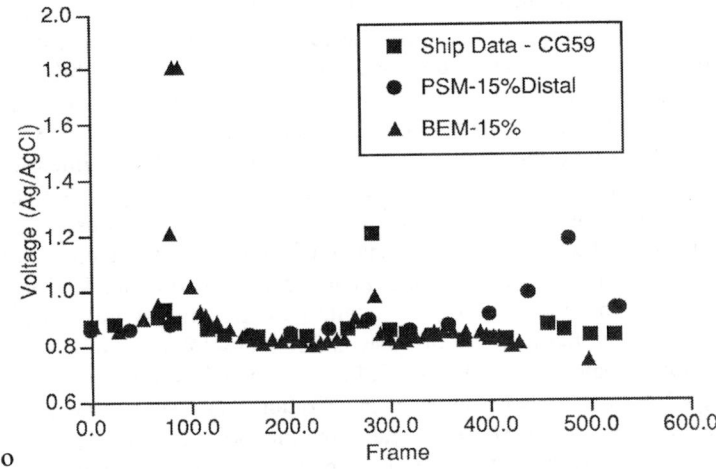

O

Figure 9: Comparison of CG hull potential profiles for real ship data, PSM and computational results (BEM) using maximum damage conditions (15%).

of the bilge keel was added to the model. This model was used in the later CG work and consists of 1583 8-noded rectangular elements (fig. 8). The elements were flat surfaces so the curved ship hull was modeled as a faceted surface. These elements were the most advanced element appropriate for use at the time. Representative results are shown in table 1 for current values and fig. 9 for potential profiles. A more detailed discussion of results is in the following section, Material Definitions.

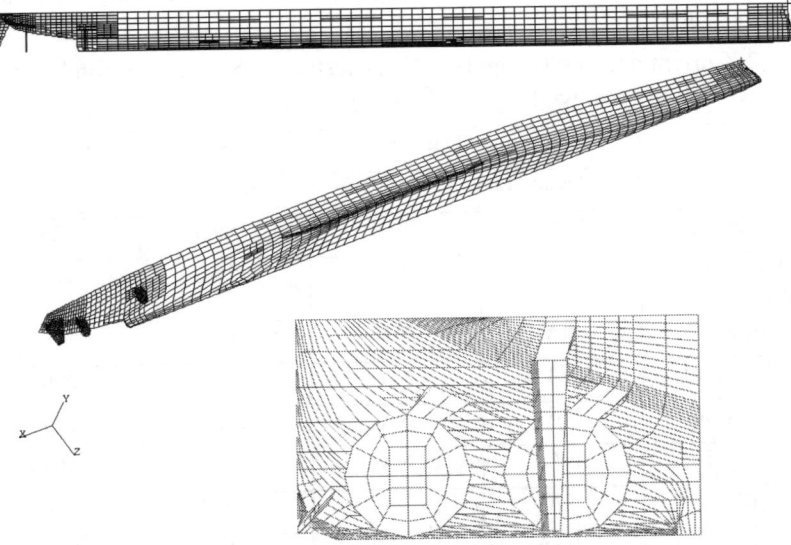

Figure 10: Boundary element mesh for CVN hull class.

The guidelines for mesh refinement and detail construction were applied when the mesh of the CVN aircraft carrier hull class was generated [29] as shown in fig. 10. Propellers were modeled as thin disks. The bilge keel was included in the mesh. In fact, the bilge keel is significantly smaller in relationship to the overall hull size as can be seen by comparing figs. 8 and 10.

The CVN boundary element mesh consists of 1884 linear-quadratic displacement 9-noded rectangular elements. The 9-node configuration consists of 8 exterior mesh points that define the element geometry and 1 mesh point placed at the centroid of the element. The centroidal node allows for curvature of the element. This element type was not available for the earlier work. The 9-noded element allows for more accurate modeling of the curved hull surface. This element was not available at the time the CG model was created. It was the most advanced appropriate element for use at the time the CVN model was created.

In the CVN analysis the source of polarization data was chosen so that PSM testing procedures would be represented by the polarization response. Typical potential contours are shown in fig. 11. Total current requirements for dynamic conditions are shown in table 2. Detailed comparisons of calculated and experimental results are presented in [29]. While potential profiles and magnitudes were accurately predicted, there was a larger degree of variation in amperage values than for the CG analysis. Possible reasons for these differences were identified as model simplification and polarization response.

All computational models are simplifications of the actual geometry. One difficult area in the CVN hull that required simplification was the bilge keel. Despite best efforts to match bilge keel profile and attachment angles there were differences between computational and PS models. These variations in bilge keel geometry are probably a contributing factor in variations observed for amperage required for mid-hull, *i.e.* bilge keel region, damaged areas. The differences in results, attributed to the variation in bilge keel geometry, highlight the need for accuracy modeling of geometric details.

Table 2: Current demand (Amps) for CVN system, dynamic flow conditions; reference cell reading = –0.85V Ag/AgCl.

	Minimum Damage		Maximum Damage	
	Calculated	PS Model	Calculated	PS Model
Propellers	118.9	201.1	189.8	228.2
Docking Blocks	71.7	110.6	185.2	181.8
Rudder	NA	NA	85.0	43.0
Bilge Keel	NA	NA	174.4	290.8
Waterline	NA	NA	206.8	229.8
Struts	NA	NA	85.8	104.4
Hull	NA	NA	791.0	759.5
Total	190.6	314.7	1718.0	1837.7

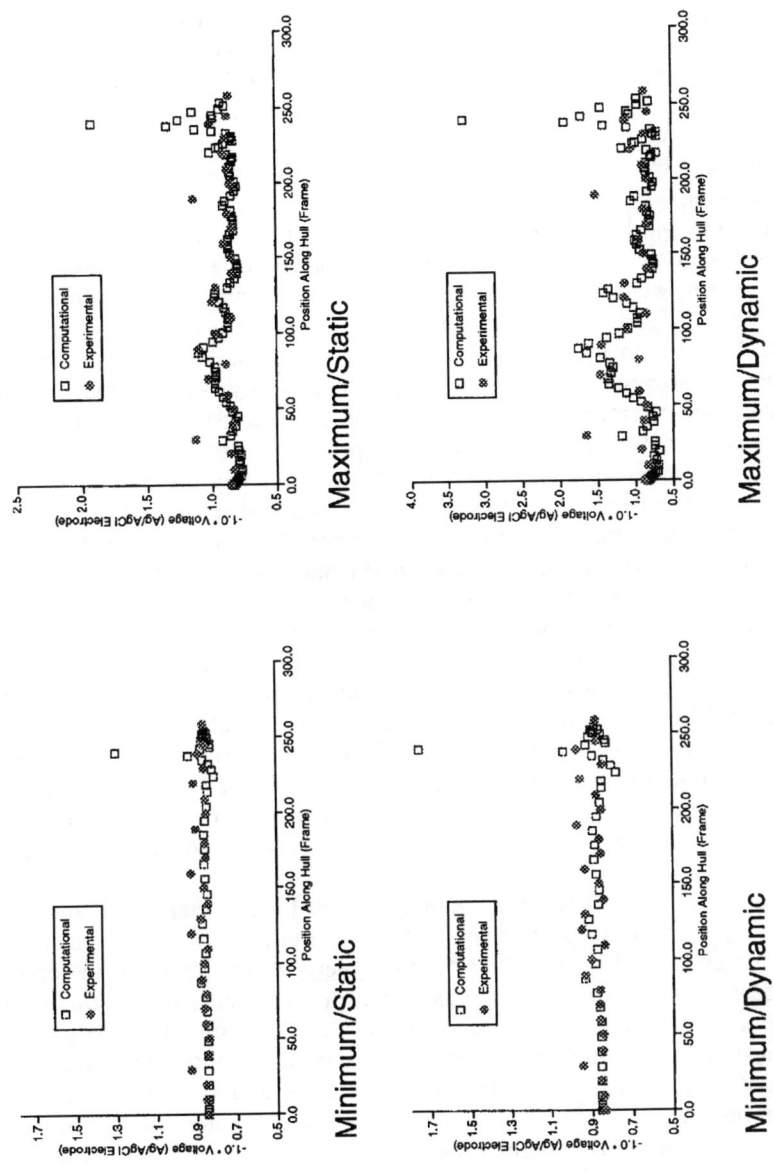

Figure 11: Potential profiles for CVN hull class; comparison of experimental (PSM) and computational (BEM) at 10ft below the waterline.

6.2 Material definitions

The CG and CVN hull meshes were both used for polarization, seawater conductivity and paint conductivity studies. Three topics will be addressed in this section; bare material polarization response, paint polarization response and seawater conductivity definitions. As noted in the real ship section of this chapter the cathodic regions of the hull are the dominant force in determining ICCP system performance. In the computational model these areas are defined geometrically by selection of elements and mathematically by polarization response data.

Once the mesh refinement study was completed for the CG hull class geometry the issue of polarization response was addressed. Initial results for the refined mesh were poor when compared with experimental measurements. Changing polarization input response to data that more accurately represented the PSM environment resulted in good agreement as shown in table 1. The change in polarization response source moved the analysis away from depending on small test specimen results from laboratory based using small specimens to data obtained from larger specimens in an open ocean environment.

A typical potential profile is shown in fig. 9 for the refined hull with bilge keel. Figure 9 shows the comparison between sea trials data, PSM experimental data and computational modeling calculated results for the USS Princeton. The comparison of experimental and calculated results determined that the accuracy of computational results is directly related to the accuracy of the input polarization data used. However, in the process of developing the refined mesh and in evaluating the impact of changing polarization input data, it was observed that performance trends were similar even when magnitudes showed poor agreement. This is an important observation since it allows for basic system design work to be performed using any reasonable polarization data. Regions of overpolarization and underpolarization can be identified quickly using standard input data curves and relatively coarse boundary element meshes. This means preliminary design work can be done quickly and for reasonable computational costs.

In the CVN analysis the polarization data used were determined from small-scale single material specimens tested in scale seawater to match the experimental environment. Despite this there were larger than expected variations between computational and experimental results for maximum paint damage-dynamic flow conditions. A review of data after the analyses were completed, indicated that there were other significant differences between the PSM test environment and the laboratory polarization experiments. While material interactions were not included in the laboratory determination of polarization response it is highly likely that these occurred due to the geometry of the hull, location of damage and location of appendages. Also the presence of film coatings, not taken into account in the laboratory polarization response, was noted on some metal surfaces of the PSM model. These variations are contributing factors to the differences in results. This evaluation again highlights the need for accurate and appropriate polarization response data as observed in the CG analyses. Polarization input data values will have a direct and dominating affect on solution accuracy.

Finally, the issue of how to represent damaged paint was addressed. It has been a convention in computational modeling to represent damaged regions of paint as regions of bare metal. In actuality damaged paint can take several forms. For instance, wear or poor application may result in thinned layers that have reduced protective capability. There may be regions where paint has been removed by scraping or other mechanical abrasive action. There may be regions where small regions of damage, holidays, may result from a variety of causes. All of these cases have been approximated by defining damaged paint regions as elements with bare metal properties.

Two parametric studies have been completed on this topic. In the first the CG hull geometry was used and the use of large areas of damage on the propeller versus the use of holidays, or small regions of damage, was examined. In the second the CVN hull geometry was used and the effect of regions of diminished paint protection, but not to the point of bare metal, was considered.

The CG hull model was used in a parametric study to determine the affect small levels of damage to the propeller region had on system performance [30]. Paint damage in this case was modeled as an effective reduction in the efficiency of the coating. The polarization curve was scaled by the effective surface area. For example, a 90% coating efficiency was defined as corresponding to 10% exposed metal. The intent in this analysis was to evaluate the effects of small areas of damage, less than the area of one element, on system performance. The effective coating efficiency is one means to model this type of damage. What was demonstrated was that the system responded synergistically. Even though changes in damage were concentrated in the aft section, both forward and aft power sources were influenced.

In a later study the impact of representing paint as a real material with degrading material properties with time was evaluated [31]. The CVN hull geometry was used to evaluate different levels of paint effectiveness associated with partial loss of paint protection but without gaps in paint coverage (holidays, or patches of removed paint). The baseline condition for this analysis was dynamic flow-minimum damage. Damaged area was added by decreasing the paint effectiveness associated with the damaged pattern used for 15% damage surface area. This results in 2.8% of the hull being defined as bare metal and in an additional 12.2% distributed over the ship hull described as intermediate between perfectly insulated and bare steel. When all damaged paint elements are assigned the polarization response of bare steel 15% of the wetted surface area is bare metal, the maximum damage condition. Input voltages values were maintained for all analyses at levels required for adequate protection at minimum damage dynamic flow conditions. Table 3 shows the changes in current draw associated with specific damaged regions. As more area is added to the damaged state, the current attracted to the original region of damage (minimum damage state) is reduced. Increased current draw is seen in the additional damaged areas as the paint effectiveness is decreased. Of note in the results is that the potential profile does not fall below protected levels until the ratio of paint polarization to

Table 3: Current total (Amps) for dynamic flow*, selected damage areas. Material properties of damaged surface varied. Damage condition varied from 3% (minimum) to 15% (maximum).

Condition	Minimum Damage		Paint Level			
	PS Model	Perfect Paint **	2	3	4	5
Paint Effectiveness ***	N/A	0	0.001	0.01	0.1	1.0
Propellers	204.1	118.9	123.8	120.0	99.9	88.4
Docking Blocks	110.6	71.7	63.8	59.3	34.5	14.9
Waterline	0	0	0.3	3.8	15.6	22.8
Bilge Keel	0	0	0.2	1.6	13.4	47.4
Struts	0	0	0.1	1.2	6.8	9.8
Rudder	0	0	0.1	0.5	6.6	10.4

*   Average current in and current out of computational results.

**   Perfect paint results as reported in [29].

*** Ratio of current demand to bare steel; 0 = insulated, 1 = bare steel.

steel polarization has reached 0.1. One interpretation of the results is that paint with a relative performance of 0.001 or less of bare steel is 'good' while any paint with a relative performance of 0.1 results in significant changes in current draw to the damaged regions. 'Good' here implies performance very similar to perfectly insulating material with minor differences in results.

There are two materials required in computational modeling: (1) polarization response for any and all different materials used and (2) the conductivity of the surrounding electrolyte. The conductivity of the seawater surrounding the ship hull has an impact on ICCP system performance. PSM has a limited range of seawater conductivity that can be evaluated due to scaling considerations. The more brackish the water, the harder it is for PSM to create the scaled conductivity seawater required for testing. There is no such limitation associated with computational modeling.

The affect that a small but realistic variation in seawater conductivity had on ICCP system performance was determined to be of interest. The refined CG mesh with bilge keel was used for the seawater conductivity analysis [32]. A twenty percent range of seawater conductivity centered on the nominal value was evaluated. This range was defined based on reported variation in seawater in different temperate zones. The analysis indicated that these moderate variations in seawater do result in moderate changes in system power requirements to maintain the set point. Of more importance was the fact that reference cell placement was shown to become a critical issue with changing seawater conductivity. Reference cell placement that provided adequate system performance at one conductivity level may or may not provide adequate system performance at a different level. Full field contours of potential levels can be obtained from the computational model and used for reference cell placement in the design process. Potential profiles at a single depth were shown to not provide a true picture of hull performance.

6.3 Boundary condition definitions

Polarization response of bare metal is a necessary material input for computational analysis. In early analysis conventional modeling practices were followed that defined painted surfaces as performing as perfect insulating materials. This does not accurately represent paint systems. While painted surfaces are much less active than bare metal regions these surfaces are not perfect insulating materials. The basic question to be answered was: Is the use of a perfectly insulating material boundary condition an adequate representation of a real material?

The definition of painted surfaces as perfect insulators, and the use of a boundary condition to identify these regions, is an assumption that simplifies the modeling process. When material definition is used all elements must be defined with a material value; this can, and did in early analyses, create problems with file sizes and program limitations. If one chooses to define paint as a material rather than use a boundary condition simplification it raises the question of what material properties to use. These values are not readily available and when exist may be proprietary. In evaluating whether it is necessary to incorporate material response for paint in the computation modeling process ranges of values are considered to identify trends. In a computational model, paints and passive coatings are similar; both passive coatings and paints result in materials that have a higher resistance level. Passive coatings have been observed to reduce conductivity values by an order of magnitude or more [33]. Paints are engineered materials and are much more effective in the reduction of conductivity, however, paints vary in their effectiveness by several orders of magnitude [34]. One rule of thumb is that paints increase metal resistance by 3 orders of magnitude. The range of paint properties presented was based on the effects of natural passive coatings and this rule of thumb.

The CVN hull mesh was used to determine the impact of modeling paint as a real material [31]. Results from previous studies are incorporated and the undamaged painted surface is assigned a polarization response scaled from that of steel. This defines a finite resistance for the undamaged painted surface. Current values were adjusted until the reference electrodes were at the target potential (−0.85V Ag/AgCl electrode). The influences of reduced effectiveness of different conceptual paint systems were evaluated. It was shown that the current to identified damaged regions increases with decreasing paint effectiveness as seen in table 4. Differences between the perfect paint and other calculated results are due to the presence of a relatively small current draw on the 85% of the ship hull defined as paint. The small draw over the distributed surface area represented by undamaged paint results in an increase in current demand to

Table 4: Current total (Amps) for 15% damage dynamic flow, selected damage areas. Material properties of damaged surface area varied.

Condition	Maximum Damage		Paint Level		
	PS Model	Perfect Paint *	3	4	5
Paint Effectiveness **	N/A	0	0.01	0.1	1.0
Propellers	228.2	189.8	136.2	160.9	282.3
Docking Blocks	181.8	185.2	3.0	29.6	178.3
Waterline	229.8	206.8	2.6	26.0	210.6
Bilge Keel	290.8	174.4	1.6	16.6	146.1
Struts	104.4	85.8	1.1	12.3	113.8
Rudder	43.0	85.0	0.9	10.0	102.7

* Perfect paint results as reported in [28].

** Ratio of current demand to bare steel; 0 = insulated, 1 = bare steel.

maintain the reference electrode target point. As the polarization response of paint becomes closer to steel, the total current required to obtain the target reference electrode value increases. Results based on the series of studies are:

- Paints with a relative polarization response of 0.1 or greater show significant differences in calculated results.
- Replacement of the perfect paint with a material that has a resistance representative of real paint results in minimal differences. 'Good' paint can be defined as that with a relative polarization of 0.001 or less of the polarization response of steel.
- Variations in calculated results that incorporate finite material behavior for painted surfaces are complex and show varying trends; however, in general, changes in calculated results for paints with a relative polarization of 0.001 or less of that of bare steel are marginal.

Most importantly for validation of established practice in the computational community:

- Paints with a relative polarization response of 0.001 or less than that of steel can be modeled as perfectly insulating materials without decreasing the validity of the computational model.

Another boundary condition that has recently been evaluated is the affect of PSM tank size on modeling results. In this case computational methods were used to determine any influence the different tank sizes, geometries and wall materials would have on data gathered experimentally. The influence of the surrounding tank wall on models of different sizes was calculated to provide a basis for scale factor selection.

The CVN mesh was used in a series of studies that evaluated the effects of tank geometry and model size on experimental results [35, 36]. This study evaluated the possible edge effects that may occur for a model of a defined size when placed in different tanks as part of the PSM process. Possible effects based on tank wall material and tank geometry were evaluated. The containment conditions are shown in fig. 12. The same ship is shown in each tank, however, a different viewing scale is used for each tank. In the work presented, three different containment geometries and two different wall material conditions per containment were evaluated. The geometry–material combinations represent not only the NRL Center for Corrosion Science and Engineering facility but also possible combinations that may be used in testing. The evaluation of these other combinations is to determine if open sea equivalent information can be experimentally obtained for the combinations of ship geometry (size), tank geometry and wall conditions defined. Identical hull geometry, material conditions for the hull and system performance levels (anode strengths) were used in all cases. Three different containment conditions were examined.

The first is the far field box, 12 by 5 by 5 ship lengths, used to represent open sea conditions. Ship lengths refer to full size ship dimensions. The second containment condition examined represents the actual test tank at the NRL Center for Corrosion Science and Engineering facility. The cylindrical tank at the NRL facility is 10m in diameter and is filled with scaled conductivity seawater for testing. The cylindrical tank is made of galvanized steel and has a 0.76mm neoprene liner. Natural seawater is scaled by the addition of fresh water to obtain the desired resistivity. Scaling factors used range from 1/40 to 1/96 depending on full size ship dimensions. In the current analysis the tank dimensions were scaled up so that the computational model makes use of a full-size ship model and full-strength conductivity seawater. The third containment condition examined is a rectangular tank similar to that used by Defense Research Establishment Atlantic Dockyards Laboratory. As in the case of the cylindrical tank, the computational model is scaled to contain a full-size ship model. The rectangular tank is 1.92 by 0.38 by 0.24 ship lengths.

The studies demonstrate that all results of interest are affected to some level by tank geometry and tank wall material. The results are important to both computational modeler and experimentalist since an understanding of the total system performance (ship hull and surrounding environment including tank geometry) is important for both methods of evaluation. It was determined that the boundary element method provided a useful tool for the experimentalist in the interpretation of measured results.

As the design team learns to rely more on computational results as independent data on the physical phenomenon associated with ICCP systems there is an increased need to quantify the impact of simplifying assumptions. This need is being addressed at NRL through a series of on-going and planned analyses. Some are computational in nature while others rely heavily on input

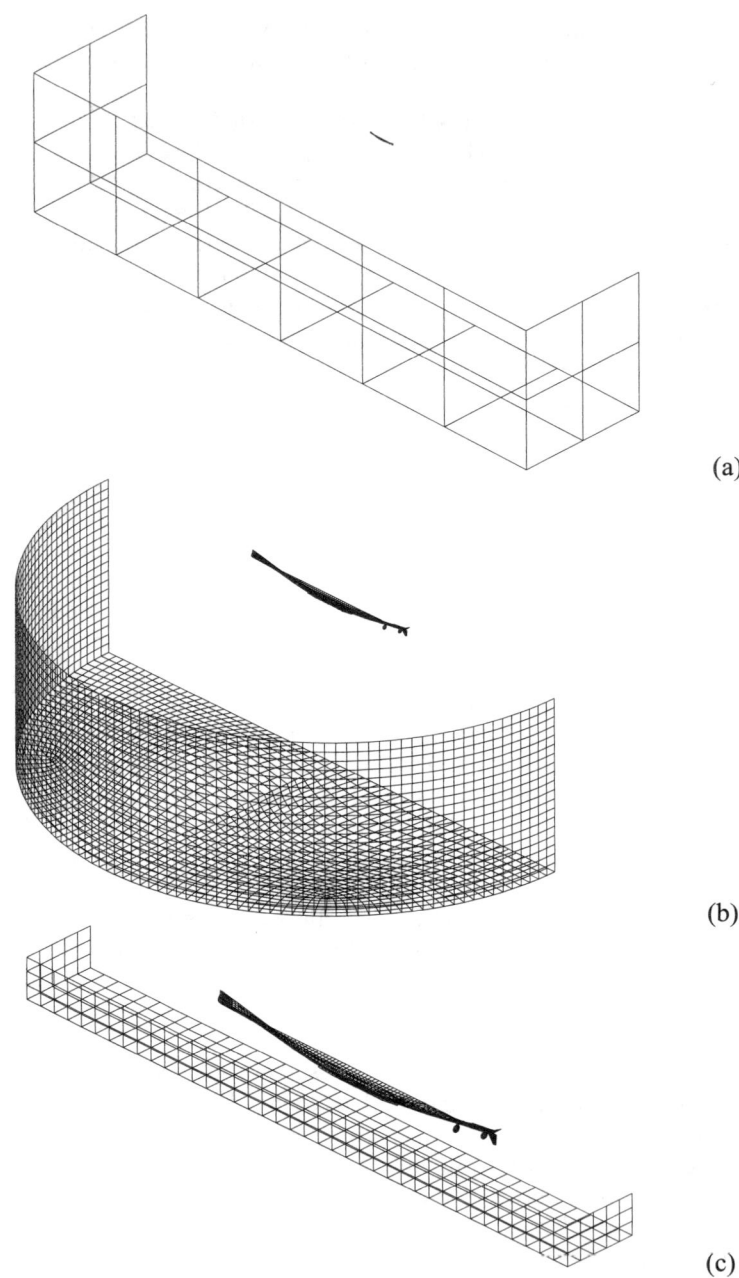

(a)

(b)

(c)

Figure 12: Computational boundary element meshes used with the CVN Hull for three different containment geometries: (a) the far field boundary, (b) the cylindrical tank and (c) a small rectangular tank.

from PSM. Topics that have been identified for evaluation include detailed modeling of propellers instead of using an idealized disk representation, evaluation of inputting the actual water surface, inclusion of free flood spaces, variations in damage patterns, littoral environment effects on system performance, seabed characteristics and ship hull symmetry. Results from these analyses will assist in the creation of a model that can best approach the complexities of the real structure. In the future, decisions on selection of model simplifications will be based on impact evaluations rather than computational size limitations or conventions.

7 The road ahead – unified design approach

Design of complex systems, such as shipboard ICCP systems, is by its very nature a multidisciplinary problem. In order to successfully apply computational or experimental methodologies to this problem the analyst has to have a basic understanding of how the system operates in the real world. Experimental and computational methodologies provide unique insights into system performance. In addition there are limitations to both experimental and computational techniques. Consequently both methods should be utilized to provide the designer with a more complete understanding of the underlying physical phenomenon. The use of experimental and computational methods in conjunction for design results in a unified design approach, shown schematically in fig. 13 [37]. The approach presented in fig. 13 involves the rapid and continual passing of information between experimental and computational approaches. In

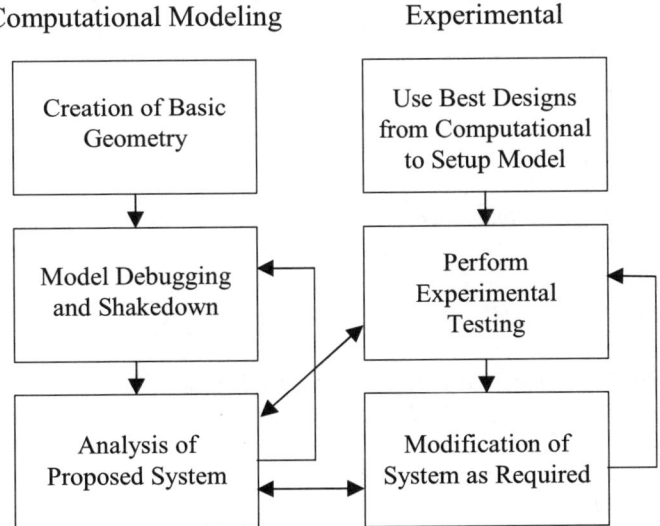

Figure 13: Unified computational and experimental design approach for shipboard ICCP systems.

the past NRL had proposed a combined design methodology that relied on both computational and experimental procedures [38] but the earlier approach was not truly integrated. Computational modeling was seen as primarily a means to reduce the number of iterations experimentation required in the design cycle. Recent advances in computational techniques and associated increased confidence in the results of computational modeling of ICCP systems has provided the basis for a full partnership between the methodologies. Computational modeling provides a means for preliminary design and can be used to establish the PSM test matrix. Key test parameters, such as model scale and relative size to tank size, can be determined computationally. In addition to linking the two methodologies it is also a means for clear and constant communication between the experimentalist and computational analyst. Each of these practitioners must realize that their particular portion of the overall design approach has its own strengths and weaknesses. For instance PSM does not have the polarization response as input data concerns that worry computational modeling. In a like manner computational modeling can provide quick evaluations of changes in system or environmental parameters that requires a much greater time experimentally. Working together the two approaches can result in a more effective design. The interchange should be seen as continual rather than a linear iterative process.

Design of shipboard ICCP systems will become more challenging in the future. Advances in hardware design, controller design, and control algorithms will increase system complexity. The next generation systems will be beyond the capability of simple design approaches. Since computational or experimental design basis are limited in their capabilities a combined unified design approach that relies on both processes is necessary. A unified design will be capable of addressing these advanced systems. On-going advances in computational and experimental methodologies will only provide greater ability to understand and apply this understanding to the design of systems.

Acknowledgements

The support of Dr Alexis Kaznoff and Mr E. Dail Thomas, Naval Sea Systems Command, is gratefully acknowledged.

References

[1] Kasper, C., *Trans. of Electrochemical Society*, **77, 78**, Papers I–IV, 1940.

[2] Agar, J.N. & Hoar, T.P., *Discussions of Faraday Society*, **1**, pp. 162–168, 1947.

[3] Waber, J.T. & Fagen, B., *Journal of Electrochemical Soc.*, **103(1)**, pp. 64–72, 1956.

[4] Ditchfield, R.W., McGrath, J.N. & Tighe-Ford, D.J., *Journal of Applied Electrochemistry*, **25**, pp. 54–60, 1995.

[5] McGrath, J.N., Tighe-Ford, D.J. & Hodgkiss, L., *Corrosion Prevention and Control*, **April**, pp. 33–36, 1985.

[6] Thomas, E.D. & Parks, A.R., *Proceedings of Corrosion 89*, Paper 274, NACE: Houston, 1989.

[7] Parks, A.R., Thomas, E.D. and Lucas, K.E., *Proceedings of Corrosion 90*, Paper 370, NACE: Houston, 1990.

[8] Lucas, K.E., Thomas, E.D., Kaznoff, A.I. & Hogan, E.A., *STP 1370*, ASTM: West Conshohocken, pp. 17–33, 1999.

[9] Adey, R.A. & Niku, S.M., *Computer Modeling in Corrosion*, STP 1154 ASTM: West Conshohocken, pp. 248–264, 1992.

[10] Munn, R.S., *Computer Modeling in Corrosion*, STP 1154, ASTM: West Conshohocken, pp. 215–228, 1992.

[11] Gartland, P.O., *et al*, *Proceedings of Corrosion 93*, Paper 522, NACE: Houston, 1993.

[12] Frazer-Nash Consultancy, *FN Remus Detailed Modeler User Guide*, Frazer-Nash: Surrey, 2001.

[13] Bardel, E.R., Johnsen, R. & Gartland, P.O., *Corrosion*, **40(12)**, pp. 628–633 1984.

[14] Adey, R.A. & Niku, S.M., *Galvanic Corrosion*, STP 978, ASTM: West Conshohocken, pp. 96–117, 1988.

[15] Brichau, F. & Deconinck, J., *Boundary Element Technology VII*, Computational Mechanics Publications: Southampton, pp. 389–404, 1992.

[16] Lee, S.H., Townley, D.W. & Eshun, K.O., *Boundary Element Technology VII*, Computational Mechanics Publications: Southampton, pp. 423–440, 1992.

[17] Yan, J.F., *et al*, *J. Electrochemical Society*, **139(7)**, pp. 1932–1936, 1992.

[18] DeGiorgi, V.G., *Boundary Element Technology IX*, Computational Mechanics Publications: Southampton, pp. 319–326, 1994.

[19] Amaya, J. & Aoki, S., *Boundary Element Technology VII*, Computational Mechanics Publications: Southampton, pp. 375–388, 1975.

[20] Gartland, P.O., Bjoernass, F. & Osvoll, H., *Corrosion 99 Research Topic Symposium CP Modeling and Experiment*, NACE: Houston, pp. 17–28, 1999.

[21] DeGiorgi, V.G., *Boundary Elements XIX*, Computational Mechanics Publications: Southampton, pp. 829–838, 1997.

[22] Diaz, E.S. & Adey, R., *Boundary Elements XXIV*, WIT Press: Southampton, pp. 475–485, 2002.

[23] Computational Mechanics, *BEASY-CP Users Manual*, Computational Mechanics Int.: Billerica, 2000.

[24] MSC Software Corp., *MSC PATRAN 2001 Users Manual*, MSC: Los Angeles, 2001.

[25] Amtec Engineering, Inc., *Tecplot User's Manual*, Amtec: Bellevue, 2001.

[26] DeGiorgi, V.G., Thomas, E.D. & Kaznoff, A.I., *STP 1154*, ASTM: West Conshohocken, pp. 265–276 (1992).

[27] DeGiorgi, V.G., *et al*, *Boundary Element Technology VII*, Computational Mechanics Publications: Southampton, pp. 405–422, 1992.

[28] DeGiorgi, V.G., Kee A. & Thomas, E.D., *Boundary Elements XV*, Computational Mechanics Publications: Southampton, pp. 679–694, 1993.

[29] DeGiorgi, V.G., Thomas, E.D. & Lucas, K.E., *Engineering Analysis with Boundary Elements*, **22**, pp. 41–49, 1998.

[30] DeGiorgi, V.G. & Hamilton, C.P., *Boundary Elements XVII*, Computational Mechanics Publications: Southampton, pp. 395–403, 1995.

[31] DeGiorgi, V.G., *Engineering Analysis with Boundary Elements*, **26(5)**, pp. 435–445, 2002.

[32] DeGiorgi, V.G., *Boundary Element Technology XII*, Computational Mechanics Publications: Southampton, pp. 475–583, 1997.

[33] Ulhig, H.H. & Revie, R.W., *Corrosion and Corrosion Control*, John Wiley and Sons: New York, 1985.

[34] Frechette, E., Compere, C. & Ghali, E., *Corrosion Science*, **33**, pp. 1067–1081, 1991.

[35] Wimmer, S.A., DeGiorgi, V.G.,Hogan, E. & Lucas, K.E., *Proceedings 2002 Tri Service Corrosion Conference*, In Press, 2003.

[36] DeGiorgi, V.G., Wimmer, S.A., Hogan, E. & Lucas, K.E., *Boundary Element Method XXII*, WIT Press: Southampton, pp. 439–447, 2002.

[37] DeGiorgi, V.G., Hogan, E., Lucas, K.E. & Wimmer, S.A., *BETECH2003*, WIT Press: Southampton, In Press, 2003.

[38] DeGiorgi, V.G., Thomas, E.D. & Lucas, K.E., *Boundary Element Technology XI*, Computational Mechanics Publications: Southampton, pp. 335–345, 1996.

Simulating the transient response of ICCP control systems

J.M.W. Baynham & R.A. Adey
Computational Mechanics BEASY, UK.

Abstract

The simulation of passive and active CP systems has over the years become a straightforward matter so that it is regularly performed to aid design and to assist understanding of system behaviour. Such numerical simulation takes into account the highly non-linear effects which take place during corrosion.

This paper describes the application of CP simulation to modelling the dynamics of the complete ICCP system, including controllers, generators, anodes and reference cells.

This ICCP controller simulation allows representation of various real effects (for example time delay in response), and allows prediction of the transient behaviour of the complete ICCP system. The ICCP system may include several controllers, and multiple anodes and reference cells.

The software, which is based on the Boundary Element Method (BEM), uses 'virtual instrument' technology to allow the user to manipulate individual controllers in the system, and to view the 'output' of the system. The 'output' can be output parameters for one or more controllers, or it can be some overall measure of response of the complete system – for example UEP signature.

Examples are given of application to multiple-controller systems.

The transient simulation runs at speeds many times faster than real-time, and so could in principle be used as part of a practical onboard control system.

1 Introduction

The simulation of ICCP systems has been used to assist design for several years [1–4]. The *classical* method used until now involves:

- ➢ definition of the output of each anode (by prescribing either voltage or current density boundary condition)
- ➢ solution to determine the resulting UEP

The way in which the anode outputs are achieved is not considered in the classical method, and so the resulting solution represents a *snapshot* of the behaviour of the complete system.

In the work reported here, attempts are made to simulate the dynamic behaviour of the control system, including the reference cell(s), the controller(s), the power source(s), and the anode(s). The resulting solutions, which could be regarded as a series of snapshots, provide a representation of the transient behaviour of the complete system.

As in the classical method, the geometry is represented using boundary elements, and the nonlinear polarization effects are represented using polarization curves, all as previously reported [5, 6].

The software is based on the commercial boundary element package BEASY.

2 Components of the system

The various components of the ICCP system are represented in the software as follows:

2.1 The power source

A power source has an input, and an output, and a description of how the output is related to the input. The input is the signal coming from a controller, and the output is passed on to the attached anode(s). The output may be a voltage, or a current. The *transfer function* which relates the input and the output may include a time delay, and may include various forms of response to a step change of input. There may be upper and lower limiting values of output.

2.2 The anode group

A group of one or more anodes is attached to a power source. The group of anodes has an input, which is the output from the power source. This group input is passed to each anode in the group, using a method which depends on the form of the input. A voltage input is passed to each anode without change, but a current input is split between the anodes using various possible methods (for example in proportions based on anode area, or in proportions which may be explicitly assigned).

2.3 The anode

Each anode has an input, which is either a voltage or a current. The output is applied as a boundary condition on the boundary element(s) which represent the anode. The transfer function which relates the anode input and output may include a time delay, and may include various forms of response to a step change of input. There may be upper and lower limiting values of output.

2.4 The reference cell

A reference cell measures voltage. A cell may sample the voltage at a single position, or it may return the average of voltages sampled at a series of positions, or it may return the area-based average of voltage sampled over an area. The input to the reference cell is obtained from the classical solution to a snapshot set of boundary conditions. The output from the reference cell is a voltage (the *process variable*, or *PV*).

2.5 The controller

A generalized controller has input values, an output value, and a method which defines how the output is derived from the inputs.

The input typically includes the *required* voltage (the *set point* or *SP*) at the associated reference cell, and measurement(s) of *achieved* voltage at the reference cell. The output is a signal (the *control variable* or *CV*) which is passed to the power supply. The method which derives the output from the various inputs can be quite general (it can be programmed in a user-defined subroutine accessed via a *dynamic link library* or *DLL*), or it can take the form of a predefined control equation. The input to the controller is not necessarily limited to the voltage measurements at its associated reference cell.

2.6 The sensor point

A *sensor point* is simply a generalized position at which voltage and electric field are measured. The input to the sensor point is obtained from the classical solution to a snapshot set of boundary conditions. The output from the sensor point is a voltage, and electric field components. The output from a sensor point may be available to a generalized controller.

2.7 The complete system

A complete system may include multiple controllers. The components attached to each separate controller (shown schematically in fig. 1) include:

> - A power source
> - A group of one or more anodes
> - A reference cell
> - Something which defines the set point

3 Simulating start-up of the ICCP system

If initially the control system is not switched-on, the UEP predicted by the software is simply the result of galvanic effects caused by any dissimilar materials of which the ship is made. Thus the reference cells and any sensor points have an input which corresponds to the classical CP solution with all anodes turned off.

When the ICCP system is switched on, each controller immediately sees its PV and SP, and sends a value of CV to its power supply. The power supply responds, and sends output to the anodes. Current flows from the anodes into the sea water, and back to the hull, producing a change in the voltage around the hull.

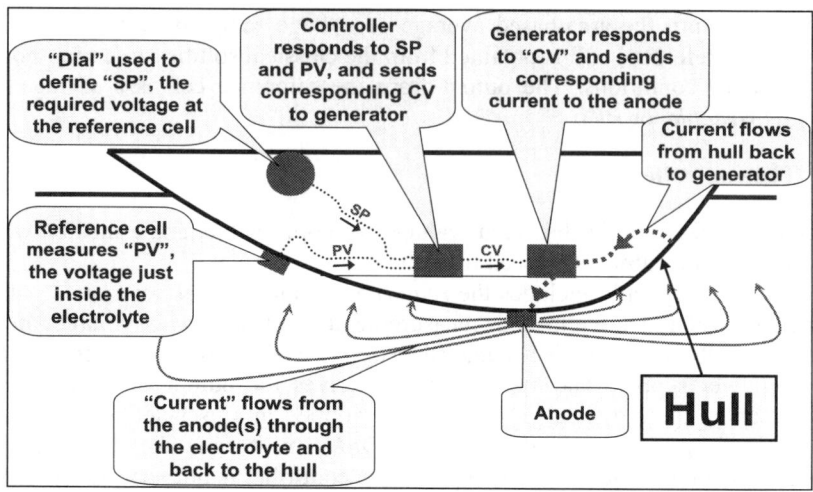

Figure 1: Schematic of a single-controller ICCP system.

The new voltage is detected by the reference cell, which passes a modified PV to the controller. The controller sees the new PV, and sends a modified value of CV to the power supply. And so on.

The mathematical description of each component is used to determine its output. The classical BEASY CP solution is used to determine the UEP (and PV) corresponding to any particular set of anode outputs.

4   The CP solution and response surface

Clearly the time taken to determine the classical CP solution from the BEM model is crucially important to the usefulness of the ICCP System simulation as thousands of solutions are required. In practise, it has been found that for realistic hull models, the time taken is too great.

Consequently, methods were developed to build a *response surface* describing the CP solution. Then, rather than performing a classical CP simulation, instead the response surface is used to determine the UEP and PV corresponding to any particular set of anode outputs. The modelling process is therefore divided into two stages, first compute the response surface from the BEM model, and then perform the simulation of the ICCP system.

Using the stored response surface method, the ICCP system simulation can operate very quickly indeed, using very limited computer resources.

Different methods can be used when constructing the response surface. The *full* response surface requires most solutions, while the *first order* response surface requires least solutions. The computational resources required to build the response surface depends upon the number of controllers, the size of the BEM model and the accuracy required.

The response surface characterizes the way the UEP varies with anode output. Notice it does *not* include any of the properties of the components (power source, anodes, reference cell or controller). This means the response surface can be used for simulations using *any* component properties.

5　The simulator as a real-time UEP calculator

Since the simulation predicts the UEP corresponding to any set of anode outputs passed to it, the software can be used simply as a UEP calculator, which can be operated either manually or automatically. When operated automatically, the controller DLL interprets the controller inputs however it likes, and returns controller outputs derived from some other source (for example measured power supply output). Since it can run at real-time speeds, the simulation provides the means to at least *monitor* the UEP.

Regardless of whether used as a real-time UEP calculator or as an ICCP System simulator, the *frequency* at which each new set of anode outputs is passed has not yet been considered in this discussion.

When functioning simply as an automatic UEP-calculator the frequency is not relevant to the calculations. The only requirement is that the software can operate at real-time. For the 4 controller system described later, and using a 1.75GHz dual processor PC, it has been found that the software can calculate UEPs at rates in excess of 1,500 per second.

When functioning as an ICCP control system simulator, the frequency may or may not be relevant to the calculations. For example an external controller DLL which represents the behaviour of an analogue controller may not use the frequency when calculating its CV output value. Whether or not frequency is used depends only on the control equation used to represent the behaviour of the controller.

In the next section, there is discussion of some control equations which *do* use the frequency when determining the controller output CV value.

6　The controller equation

So far the way in which the controller works has not been discussed. Now, however, we give some examples of control equations which may be used ...

Each of these equations assumes that a new set of PV values is available at a frequency f, which may for example be 50Hz, but could take any value. In the following:

- ➤ values which relate to a time $t - 2\Delta t$ are labelled using subscript $k - 2$
- ➤ values which relate to a time $t - \Delta t$ are labelled using subscript $k - 1$
- ➤ values which relate to a time t are labelled using subscript k
- ➤ values which relate to a time $t + \Delta t$ are labelled using subscript $k + 1$
- ➤ and so on

Note that the sampling interval T appears in the equations shown below, and that $T = 1/f$.

6.1 1DOF PID

A 1 degree of freedom proportional integral differential (1DOF PID) control equation calculates the new controller output (CVk) using:

$$CVk = CV(k–1) + Pk + Ik + Dk$$

where the values of Pk, Ik, and Dk are determined as follows:

$$Pk = –Kp(PVk – PV(k–1))$$
$$Ik = KiTek$$
$$Dk = –(Kd/T)(PVk–2PV(k–1) + PV(k–2))$$

and:

PV = ref cell voltage
T = sampling time interval
e = SP–PV
SP = required voltage
Kp, Ki, Kd = controller 'constants'
k,(k–1), (k–2) = step numbers

6.2 2DOF PID and trapezoidal PID

The so-called *2 degrees of freedom* PID control equations and the *trapezoidal* PID control equations do basically the same thing as the 1DOF, but in modified forms.

6.3 External controller

The *external* controller is exactly that – completely unknown to the software. This is because the only interaction between the ICCP System simulation software and the external controller is via a user-defined subroutine which is contained in a dynamic link library (DLL). The external controller receives input data, processes it somehow, and passes its output data back to the simulation software.

Thus the external controller can take any form at all.

7   Example: ship hull with 4 independent controllers

A *very* idealized ship hull (see figure 2) with a bronze propeller has 4 independent controllers. The anodes associated with each controller are shown in fig. 3. Details are as follows:

> The controllers used here are all 2DOF PID, with controller constants initially set at:
> > Kp = 2.4
> > Ki = 25.0
> > Kd = 0.01

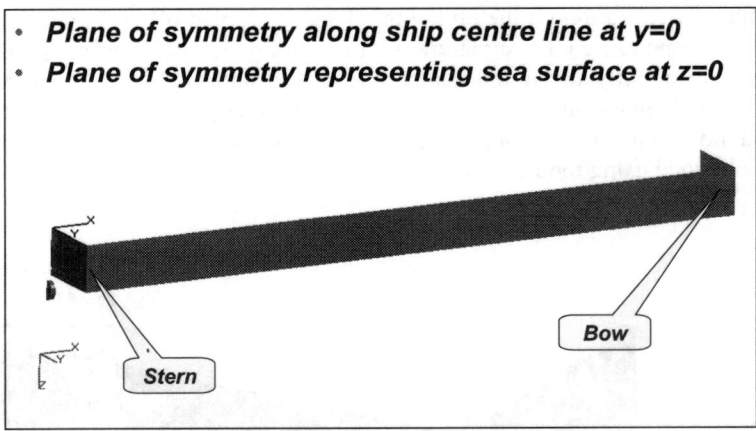

Figure 2: Geometry of hull and propeller.

> The sampling rate for the entire system is 50Hz. The transfer characteristics for the anodes relate output after unit step change of input at time t = 0 using:

Anode output (at t < 0.1) = 0

Anode output (at t > 0.1) = 1–0.5e-(t-0.1)

> The set point is – 850mV for all controllers.
> The hull coating is assumed to leak 0.1% of the bare-steel current density at any potential difference.
> There are 100 sensor points arranged in a line below the centre line of the ship hull.
> The power supply output is limited to the range 0 to 500 amperes.

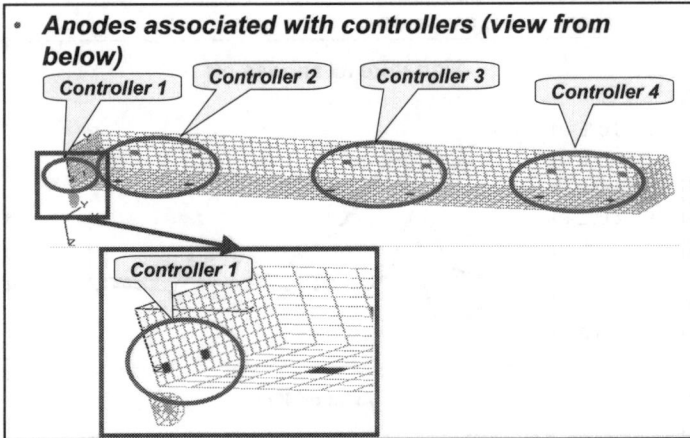

Figure 3: Showing anode positions and anode groups associated with each controller.

The voltage distribution on the hull with all anodes turned off is shown in fig. 4, and the corresponding UEP signature is shown in fig. 5. The sampling points are along a line directly under the centre line of the hull. These plots (of results which were obtained using the classical method with zero current density applied to the anode elements) are presented so that comparison can be made with the results obtained using the ICCP simulator.

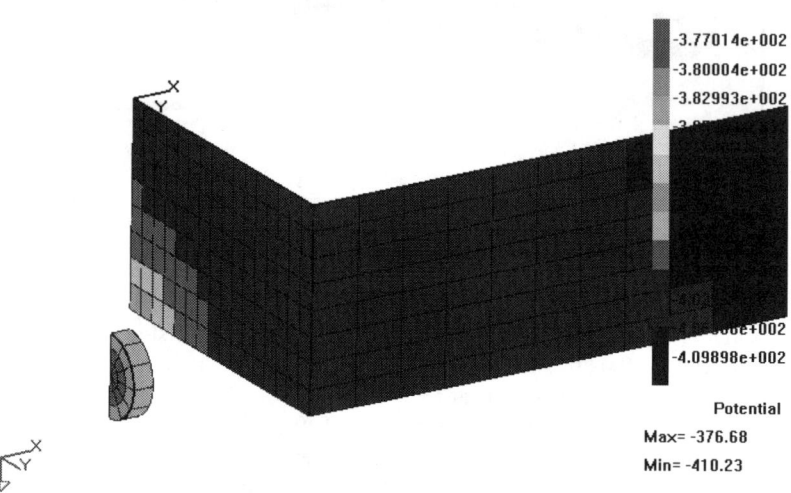

Figure 4: Potentials with all anodes turned off.

- *All anodes OFF*

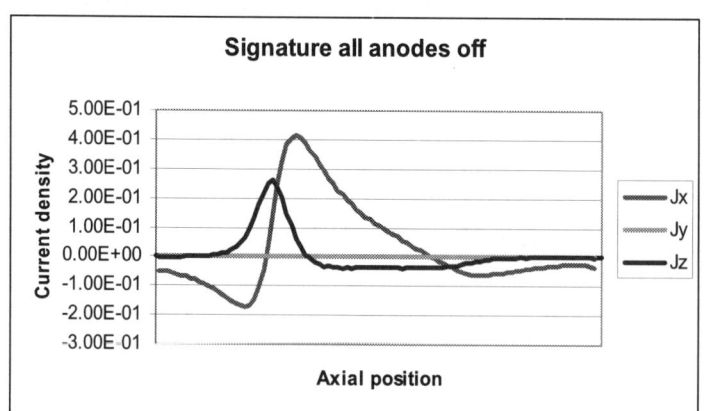

Figure 5:   Signature with all anodes turned off (classical method *i.e.* direct from the BEM model).

Using the ICCP simulator, first the response surface is read, together with details of the controllers, anodes, transfer function characteristics, and so on. The transient simulation can then be started, with real-time graphing if required. With the controllers all set to manual, and the control variable set to zero, the ICCP simulator should replicate the signature obtained using the classical method, and indeed it does, as can be seen in fig. 6.

Next the controllers are all switched on, and the system is allowed to reach a steady state. The steady state UEP predicted by the simulator is shown in fig. 7.

- ***Signature when all anodes are off, from ICCP Sim***

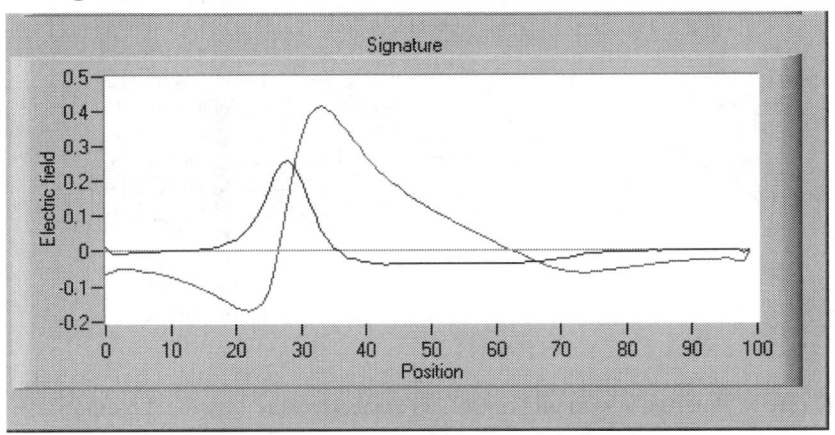

Figure 6: Signature with all anodes turned off (ICCP Sim method).

- ***Signature when all anodes reach steady state, from ICCP Sim***

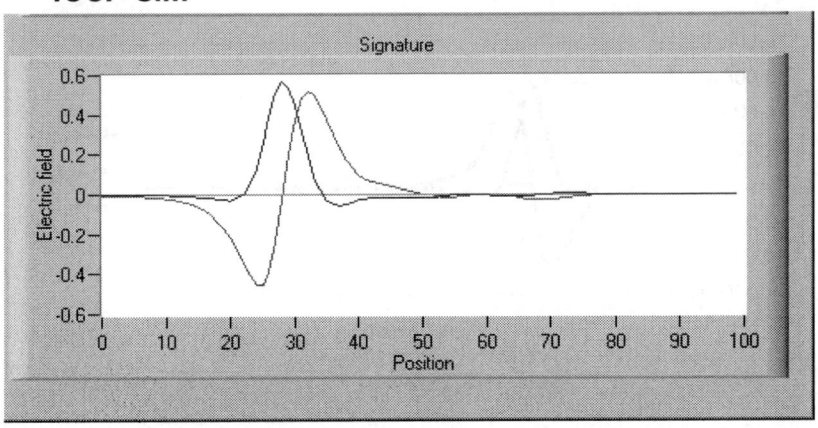

Figure 7: Signature with all controllers at steady state (ICCP Sim method).

As a final test, the steady state anode currents from the transient simulation have been applied using the classical method, and the resulting voltage distribution on the hull and UEP signature are shown in fig. 8 and fig. 9.

• *All controllers at steady state*

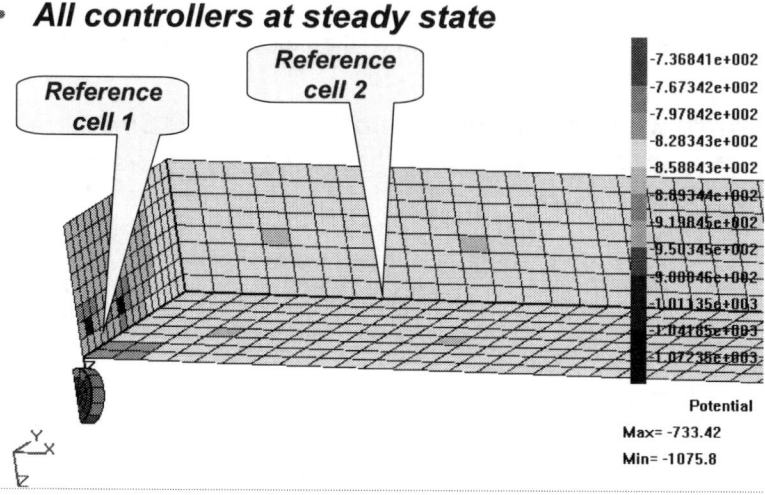

Figure 8: Potentials with all controllers at steady state (classical method).

• *All controllers at steady state*

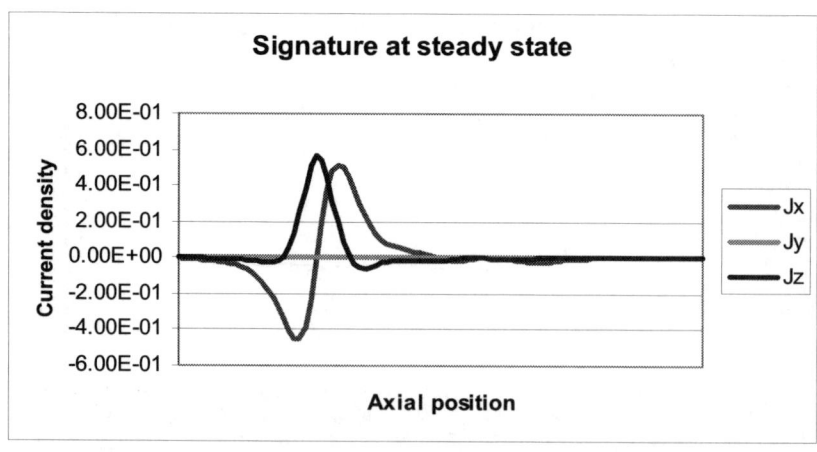

Figure 9: Signature with all controllers at steady state (classical method).

8 Discussion

Comparisons of fig. 5 with fig. 6, and of fig. 7 with fig. 9, show that the UEP signature calculated using the ICCP Sim method is the same as the signature calculated using the classical method. This is the expected result.

The correspondence occurs naturally in the case where all anodes are turned off, since the response surface was constructed with '*all anodes turned off*' as one of the combinations of anode outputs.

In the case where the anodes are at steady state after initial turn-on, the anode output values are *not* at values used during construction of the response surface. Hence this case confirms that ICCP Sim is producing the right answers.

8.1 The effect of the component properties

The time-delay and transient response of the various components are important to the dynamic simulations. Investigations to determine the form of transfer function most appropriate to a ship design may require calibration using experimental measurements.

9 Applications

9.1 Controller tuning

One of the applications of the software is to assist in selecting the controller constants Kp, Ki and Kd for a digital controller. The values of the constants have an enormous effect on the way in which the control system responds when the ICCP system is switched on, and the result of poorly chosen values may be either very sluggish response, or at the other extreme an unstable response.

Using the controller constants described earlier, the transient response for each of controllers 1 to 4 is shown in figure 10.

- *Transient potentials at reference cells after switch-on, using optimal constants*

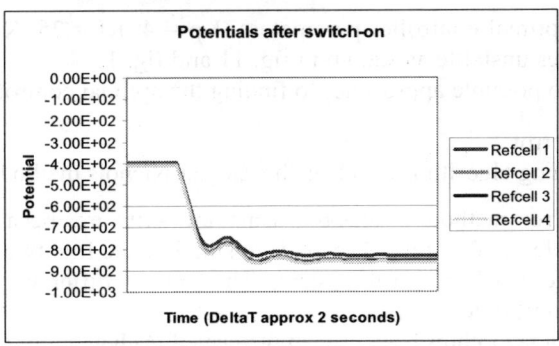

Figure 10: Transient response after switch-on, using 'optimal' constants.

- *Transient potentials at reference cells after switch-on, using non-optimal constants*

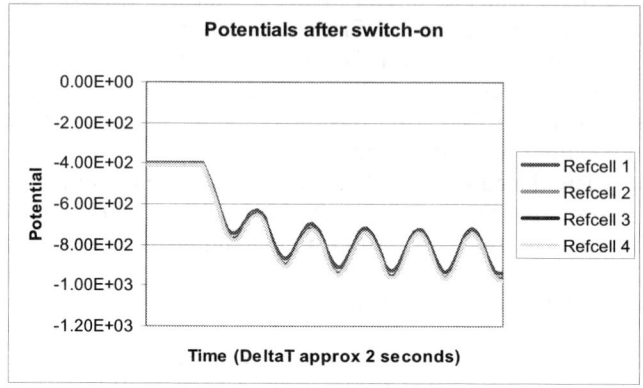

Figure 11: Transient response after switch-on, using 'non-optimal' constants.

- *Transient potentials at reference cells after switch-on, using non-optimal constants*

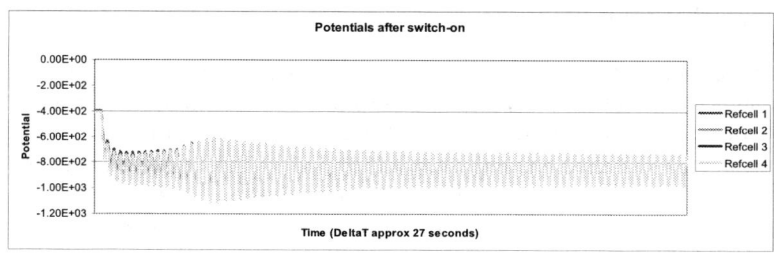

Figure 12: Transient response after switch-on, using 'non-optimal' constants.

Using non-optimal controller parameters (Kp = 4, Ki = 25, Kd = 0.01) the response becomes unstable as shown in fig. 11 and fig. 12.

There are two possible approaches to finding the optimal controller constants:

➢ trial and error
➢ use of tuning algorithms, such as the Ziegler Nichols methods

The trial and error method is feasible using the software, because for the 4 controller example, it takes only 4 seconds to simulate the first two minutes after switch-on of the ICCP Control system. Thus many different trials can be performed in a short time.

Various tuning algorithms have been implemented. Although the Ziegler-Nichols open loop reaction rate method was used to find the controller constants earlier reported as 'optimal', it is not yet clear that these methods are entirely suitable.

9.2 Controller interaction effects

Interactions may occur between multiple controllers, and investigation of such effects is necessary to understand the real cause.

A good example of interactions is given by the 4 controller example, which from fig. 10 appears to have achieved the design aim after only 1 second. However, there is in fact an interaction taking place, as can be seen from fig. 13 and fig. 14, which show variation with time of the current delivered by each controller.

- *Transient total current for controllers after switch-on, using optimal constants*

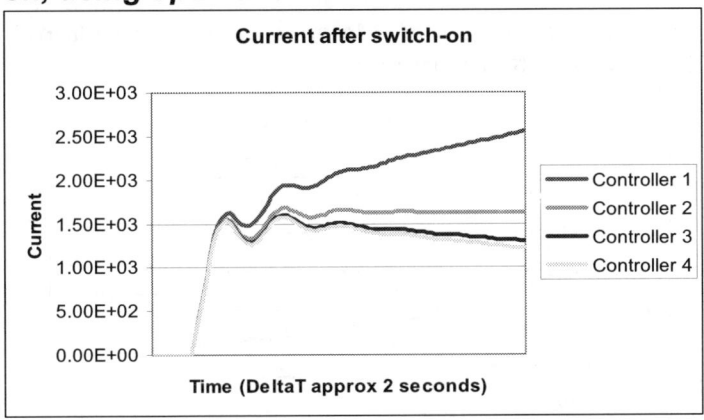

Figure 13: Variation of current for ~2 seconds after switch-on.

- *Transient total current for controllers after switch-on, using optimal constants*

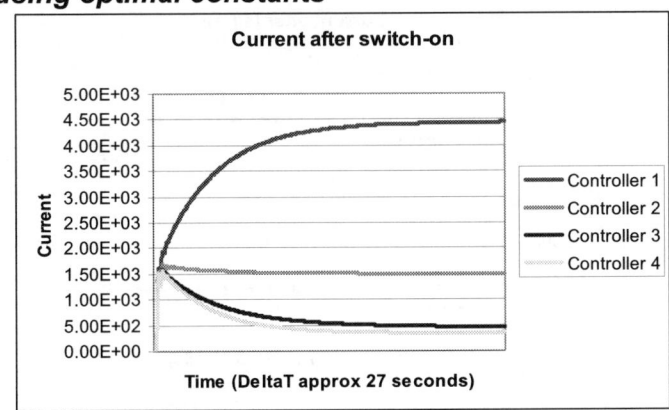

Figure 14: Variation of current for 27 seconds after switch-on.

It is clear that the effect of increasing controller 1 current is partly negated by simultaneous decrease of current by controllers 3 and 4. Closer inspection of the reference cell potentials (see fig. 15) clarifies what is going on: in fact the reference cell 1 potential has not reached the set point even after 4 seconds.

9.3 Simulating controller failure

The dynamic simulations can very easily represent the effect of failure during operation of one or more controllers. To achieve this, the selected controllers are turned off during the simulation and their CV is set to zero. The effect of failure of controller 2 is shown in fig. 16 and fig. 17.

* *Transient potentials at reference cells after switch-on, using optimal constants*

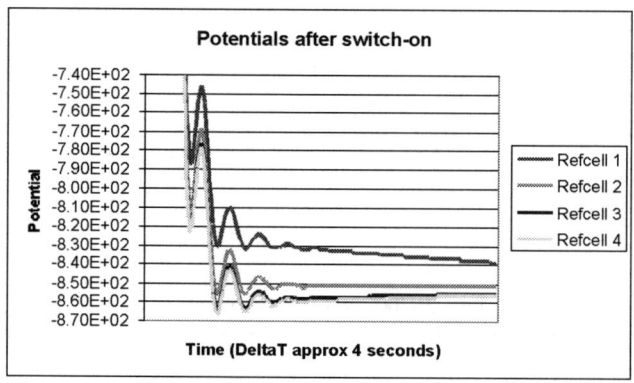

Figure 15: A closer look at reference cell voltages after switch-on.

* *Failure of controller 2*

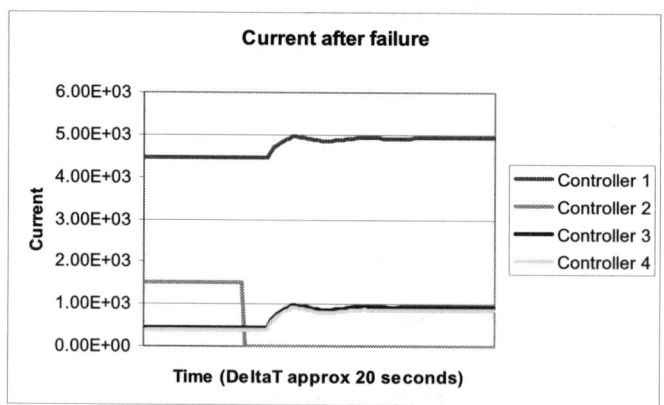

Figure 16: Effect of failure of controller 2 on controller current.

- ## *Signature after failure of controller 2*

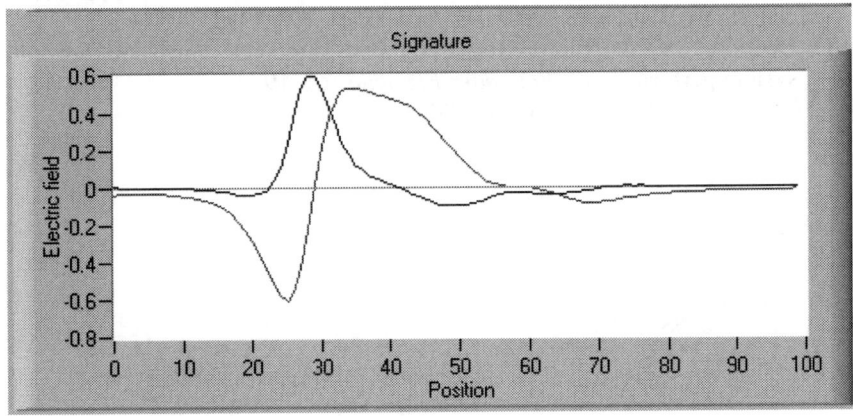

Figure 17: Signature after failure of controller 2.

10 Conclusions

The work reported in this paper shows how a BEM model of the electric fields and corrosion electrochemistry coupled to a model of the ICCP control systems can be used to represent the transient dynamics of the system.

Simulation of the dynamics of the control system allows study of start-up behaviour, interactions between controllers, stability, and other effects.

The methodology can be used simply to predict the electric signature for any given anode outputs, and the speed of the method makes it possible to monitor the changing electric signature in real-time using very limited computing resources.

Acknowledgement

Thanks to NSWC for their input during development of the software.

References

[1]   Adey, R.A., Brebbia, C.A. & Niku, S.M, 'Applications of Boundary Elements in corrosion engineering', *Topics in Boundary Element Research*, Computational Mechanics Publications 1990.

[2]   DeGiorgi, V.G., Thomas, E.D & Kaznoff, A.I., 'Numerical Simulation of impressed current cathodic protection systems', *Computer Modelling for Corrosion*, ASTM STP 1154, 1991.

[3]   Trevelyan, J. & Hack, H.P., 'Analysis of stray current corrosion problems using boundary elements', *Boundary Element Technology IX*, 1994.

[4]   DeGiorgi, V.G., Lucas, K.E., Thomas II, E.D. & Shimko, M.J., 'Boundary Element Evaluation of ICCP Systems Under Simulated Service Conditions', *Boundary Element Technology VII*, 1992.

[5]   DeGiorgi, V.G., 'Finite Resistivity and shipboard corrosion prevention system performance', *Boundary Elements XX*, 1998.

[6]   The BEASY CP User's Guide, 2004.

Boundary element modelling of complex multi-hull surfaces of hypothetical future warships

A.S. Christopoulos
Defence Science and Technology Organisation, Australia

Abstract

BEASY CP (Version 8.1) has been found to be an extremely useful computational tool for calculating both corrosion-related surface potentials and underwater electrostatic fields or UEP signatures of warships. In this review paper we have created and examined an hypothetical model of a prospective high-speed amphibious lift craft. The paper sets out for the reader the means of how we optimise the boundary element software to both perform essential modelling and extract key data relating to surface potential and UEP signature.

1 Introduction

This is a review article which intends to cover or highlight some of the practical applications of the boundary element technique, particularly when at this point of time, we recognise an acute requirement for rapid, accurate and innovative advice for our respective military directorates.

As will be demonstrated, the boundary element method (BEM) is a very useful tool for the theoretical examination of corrosion patterns and underwater electric potential (UEP or static electric) signatures of ships and submarines. Therefore, the review, as with many other of my past papers, will focus on a hypothetical naval platform; a catamaran system. The operational feasibility of a high-speed catamaran for the purpose of providing fast amphibious lift, that is, conveying troops rapidly from ship to shore, was investigated in 1999–2001 by the Royal Australian Navy (RAN) through the deployment of HMAS *Jervis Bay*. During its two-year service period, this platform showed to be a very efficient and effective system.

It must be pointed out however, that the model developed and theoretically examined in this review is as stated, purely hypothetical. It does indeed represent a two-hull system as per HMAS *Jervis Bay*, but the dimensions, hull curvature, material distribution propeller system and so forth do not correlate. That is to say, the model may, by remote and unforeseen chance, only bear a vague resemblance to another catamaran system, but it is in no way a simulation of any existing or operational naval platform.

The software package implemented is BEASY CP Version 8.1, released by Computational Mechanics BEASY, Southampton, UK. I have used this and earlier versions to analyse and optimise both the cathodic protection (CP) and UEP signature of a number of RAN and Royal Navy (RN) platforms very much to the satisfaction of the customer, for approximately 4 to 5 years.

The next section(s) will briefly discuss the boundary element method as well as the finite element method, both valuable computational techniques, proceed with the discussion of the geometry, meshing, material assignment and environmental conditions used to model the catamaran, whilst the following sections will place greater emphasis on how the active anode is replicated, implemented and optimised to provide both adequate cathodic protection and minimise the UEP signature.

2 The geometry and modelling of the catamaran

2.1 The Boundary Element Method and the Finite Element Method

The Boundary Element Method, (BEM), could be described as a supplement, a competitor and in due time, perhaps even a replacement to the Finite Element Method, (FEM). The FEM, another valuable numerical technique, has been used by a number of defence research agencies as well as Australia mainly to determine the induced magnetisation of warships as to effect of their degaussing coils. The technique demands that the entire volume of the model, extending to the pseudo-infinite boundaries, are meshed with continuous 3D elements. This requires an extensively large number of elements to build a model of the vessel that shall hold a feasible level of precision. Indeed, the ratio of elements with respect to BEM:FEM when constructing an accurate model would be approximately 1:30, this ratio determined after conducting many computations via each methodology. The advantage of the complete volume meshing in the case of warship degaussing is that a hull thickness and ferrous internal components of the warship can also be designed into the model. In addition, an exact geometry and positioning of the degaussing coils relative to the inner hull is obtained. The internal components such as the bulkheads and motors can themselves be affected by the degaussing coils and therefore present a further perturbation to the induced hull magnetisation. However, because of the need for element continuity, the geometry of the hull and internal components can become extremely difficult to model with warships. The ensuing approximations thereby have a tendency to force unwanted systematic errors, which in turn impede the accuracy of the result.

In contrast, although the BEM is not as capable of obtaining the numerical solution to as great a variety of physical problems as the FEM, its modelling technique is simpler to use and importantly, it therefore has a better accuracy. The exception to this is the calculation of non-linear problems such as monopoles, where the FEM has been found to generate results of greater accuracy. As noted by Trevelyan [1], this can be attributed to the larger matrix used via the FEM, traditionally known as the 'stiffness' matrix, to converge with greater reliability. A greater reliability in convergence implies that the matrix is more easily reduced, and we do not face the difficulties of ill-conditioned matrices, which, under most circumstances, cannot be reduced and thereby generate an acute divergence.

The Boundary Element and Finite Element Methods are, currently, best used as complementary techniques, rather than being placed in opposition to one another. It would be prudent for the research scientist to be proficient in both the recognition and application of which technique would be the most profitable in solution of complicated electric or magnetic warship problems.

2.2 Geometry definition

We wish to simulate the surfaces of our hypothetical catamaran below the water level or its two underwater hulls along with their appendages through the BEASY methodology. The principal (and personally preferred) source of information available is the ship's drawings or more specifically, its line plans which essentially designate the shape of the hull and the shape and positioning of other external components such as the propellers, propeller shafts, rudders and so forth. Some modellers prefer to import CAD file models. Irrespective of preference, in the case of the catamaran hulls for which no line drawings or CAD models exist, they were modelled according to the author's own experience and discretion, as discussed in the next paragraph. The water line length of the catamaran system is 100m, each hull identical with a maximum width of 7m. The centre line of each hull is 7.5m from the x-axis, in contrast to a monohull system, where the hull centre or keel line is along the x-axis. The lowest point along the keel line is 5m below the water surface, or at $z = -5$m. The total beam width at water level, is 22m.

A hull surface is created by a series of 20 parabolic functions (that is, a simple $y^2 = -z$ relation at 20 discrete values for x). Their size, shape and position with respect to $\{y, z\}$ in order to create a feasible hull surface. On completion of the hull, a single drive shaft, its support and a rudder are mapped into the model, their dimensions (suitably) estimated relative to the created hull. A propeller's surface area was estimated and was hence added to the model as a two-blade system of comparable surface area in order to reflect an accurate simulation of the distribution of the electrolyte in contact with Nickel Aluminium Bronze (NAB). The fact that there is symmetry about the x-axis, (aside from the propeller), can be taken advantage of, assuming that the surface conditions of each hull are characteristic to those of its reflection. Figure 1 displays an (aft) end-on view of a single hull. This model is hence translated and reflected across the x-axis, after

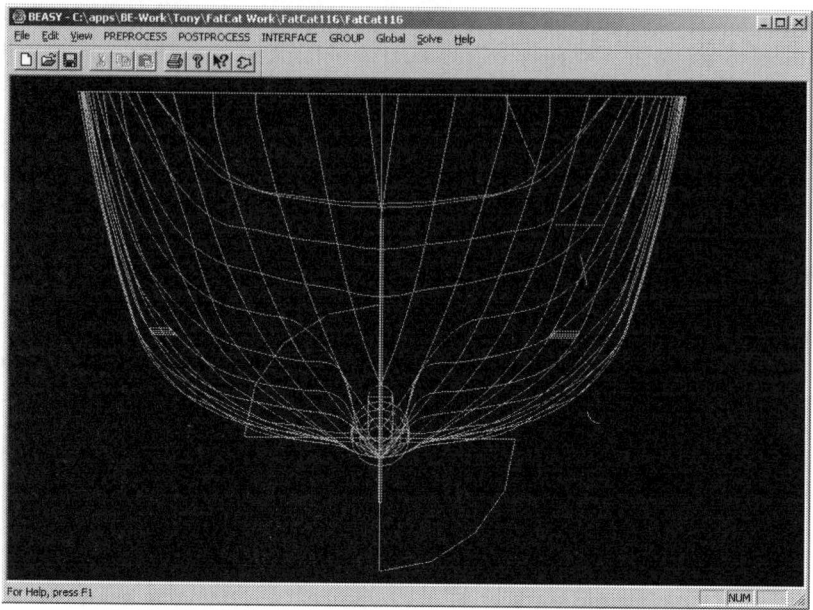

Figure 1: (Aft) end view of hull geometry.

patching and meshing, (see Section 2.3 below), to create our catamaran surface.

The pseudo-infinity boundary conditions (discussed below) are assigned to the surfaces of the outer 'box'. This is required to have limits at least 20 times the maximum dimensions of the catamaran or indeed, whichever vessel or structure it contains. The length, width and depth of the outer box for this are 2,000m, 500m and 150m respectively. In addition to a complete wet end of the catamaran model, and the pseudo-infinite boundaries, the 'box walls', is a plane parallel to the water surface and located at $z = -21$m. The plane is 21m below the water surface and 1m below the 'sensors' or internal points. This is a simulation of the water/seabed interface and hence known as the interface plane. The patching and meshing of the interface plane proves to be of crucial importance if we are to expect any accuracy from our model(s) when computing the static electric signature. The inclusion of the interface plane provides a very realistic system with respect to the immediate ranging environment. The topic will be discussed in detail in later subsections, where we introduce the zone concept into our model.

2.3 Patching and creating the element mesh

Patches are finite segments of the surfaces of the model, which specify the side of the surface where the electrolyte exists. A patch is created by three (or four)

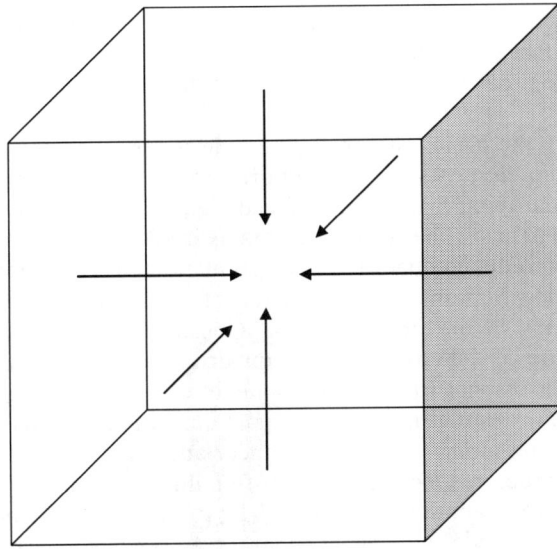

Figure 2: Patching and identifying the electrolyte's location.

conjoined lines, arcs or splines forming its boundaries in a specific order. The simplest way to determine where the electrolyte exists and how the patches are to be created is to visualise an elementary 3D object, a cube, say. In this example, consider six arrows originating from the centre-point of a cube with the condition that they all must point *away* from the electrolyte. Extending our imaginations further and accepting the cube as our warship, the electrolyte would be *outside* the cube, and therefore, the arrows must be directed *toward* the centre-point, as shown in fig. 2. Returning to the catamaran and the lines which is to form its patches, the patch and hence the electrolyte's location is modelled by progressively identifying its lines in a clockwise direction, when viewing the model from *outside* the vessel. In contrast, when viewing the inner surface of the model, the lines are taken in an anti-clockwise direction to form the required patch. By observing where the viewpoint is with respect to our model and following this procedure, the electrolyte, as realistically expected, will be set to lie outside the two hulls. Similarly, the rudders, the drive shafts and their supports, and propellers are patched. Note that as all surfaces are to be meshed, the walls representing our pseudo-infinity and the interface plane are also patched and hence meshed. The pseudo-infinity walls require no essential patching format as to their shape or distribution other than the fact that the electrolyte will be lying within the walls and that the interface plane comes in contact with them. The patching of the interface plane can be a little more demanding as the accuracy of the model, or rather the static electric signature we wish to predict from it, heavily relies upon it. This is discussed in greater depth in the coming paragraphs. Another point to note is that patching must be conducted prudently and with good precision when creating the much smaller surface regions that represent anodes. In general however, surface patching is a relatively simple procedure for the majority of the model and can be completed quite rapidly.

Elements are the basic units that subdivide the patches in the construction of a model. The elements are not required to be universally continuous on every surface, implying that different sized elements may be used in different and adjoining patches. The types of elements used in the BEASY code range from constant to reduced quadratic. Different element types are indicative of the nodes or mesh points characteristic of each type, and therefore of potential accuracy of the final solution. Constant elements having a single node at the centre of the corresponding quadrilateral or triangle, are the most economical with respect to computational time and provide good accuracy, especially when calculating surface potential distributions or assessing cathodic protection alone. The use of linear and quadratic elements, having four and nine nodes respectively, may offer improved precision, but at the expense of lengthy computational time. The use of the constant element will in theory also provide a rapid convergence with respect to the given polarisation data.

In the case of the catamaran, the constant element is appropriate to use entirely for the hull surface and the surfaces of its drive shaft supports, drive shafts, propellers and rudder. Meshing of the catamaran model is illustrated through figs 3 and 4. Similarly, the pseudo-infinite 'box' walls are adequately meshed via constant elements. It is not necessary to conduct an elaborate meshing at the pseudo-infinity, walls because their only distinct purpose is to define a zero current density, or electric flux density of zero, at a great distance from the hull(s) or surface of interest.

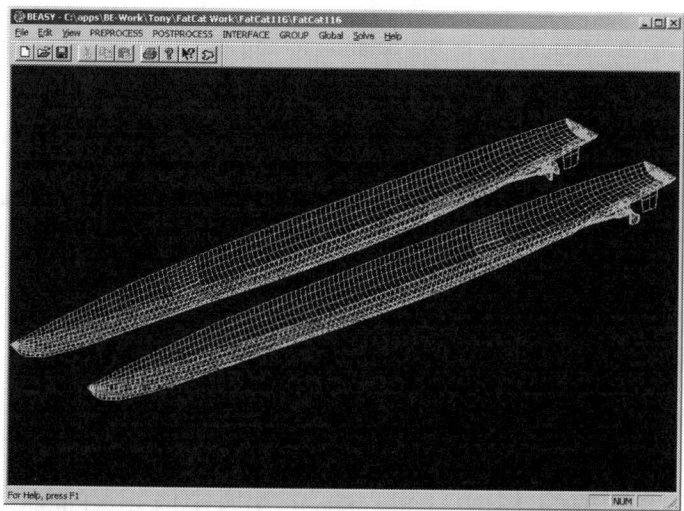

Figure 3(a): Meshed catamaran, hulls and appendages only.

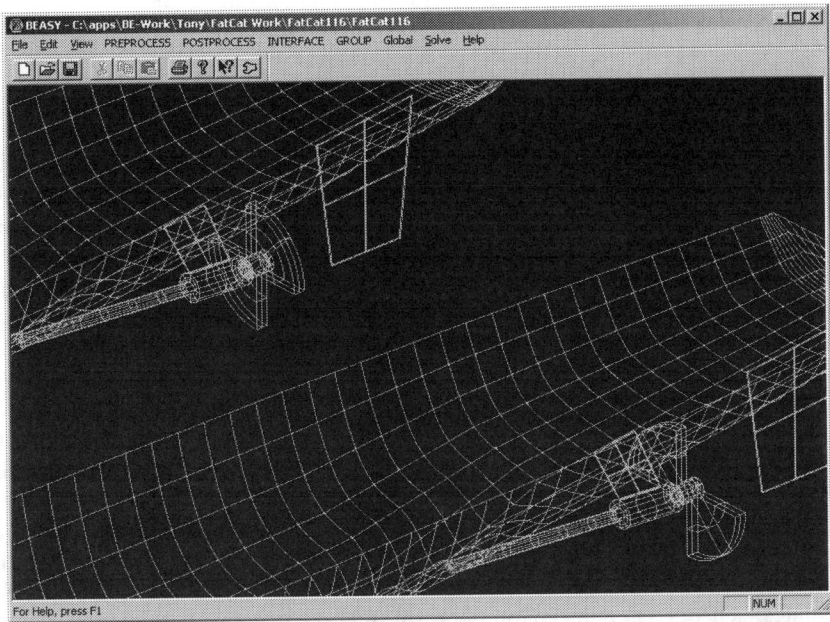

Figure 3(b): As per fig. 3(a), but zooming in to observe stern appendages.

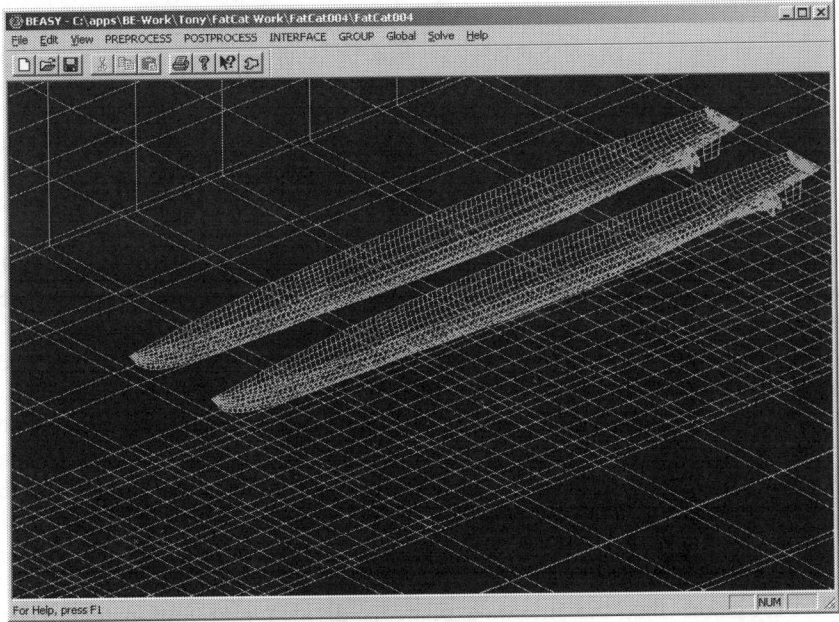

Figure 4:   Meshing of catamaran model, including view of seabed element distribution and internal points.

2.4 Water/sea bed interface elements and internal points

The surface, which demands much attention, is water/seabed interface plane. In order to replicate a realistic range environment throughout the model, it is essential that this interface plane is included. If it were not, and thereby assuming an infinite water depth, we would also be assuming that the conductivity of water remains constant. This is not physically realistic as the conductivity of the seabed is not the same as that of water. Therefore, we are implying that the electric flux density as induced by the seabed is ignored, and the computation of the electric signature would be no more than an empirical estimate. It would still be confirmed that our computation of the surface hull potential remains correct and unaffected, whether or not the interface plane was present because the ship is in contact with water only. (In contrast, computation of induced magnetic signatures via the FEM is indifferent to non-ferrous materials underwater because, as that of air, their relative permeabilities, are unity, and such materials are therefore magnetically transparent).

Signatures are of course, neither calculated nor measured on hull surfaces. If, for example, the multi-influence range sensors are positioned Zm below the water surface, then the sea bed would be $(Z + 1)$m below the surface. BEASY simulates sensors as a large number of internal points or reference points. Internal points are used to extract information such as potential and current, and 50 internal points can be used along a line with length of L_{IP} directly below the keel, and any point on this line is defined as

$$\{x : x' \le x \le x'', y = 0, z = -20\}\,\text{m} \tag{1}$$

where x' is 900m, and x'' is 1,100m, implying that L_{IP} is twice the water line length of a catamaran hull. Up to 50 internal points are required to ensure that the outcome of the calculated signature is undertaking a relatively smooth format.

Returning to the interface plane and recalling that it is to lie 1m below the internal points, we must now consider the level of detail required for the meshing. Trevelyan [1] states that '…from the BEASY software, version 5.0 recommends that the internal points be no closer than 15% of the element side length for a 3D problem.' If d_{IP}, the perpendicular distance of an internal point from the interface plane, is 1m, and d_E is the length of a (square) element's side, then this statement implies that $d_{IP} > 0.15d_E$, or more directly, that elements can have a length no more than six times the perpendicular distance between the internal point and the interface plane. Although this is conceptually sound, it does not apply to the current BEASY version. It has in fact been found from past work [2–4] that if d_{IP} is 1m, then d_E must be less than 4m.

Let's assume that the interface plane is divided into four patches, each 1,000m in length by 250m in width, then each patch would demand 256 × 64 (square) elements for d_E to satisfy this condition. The total number of elements

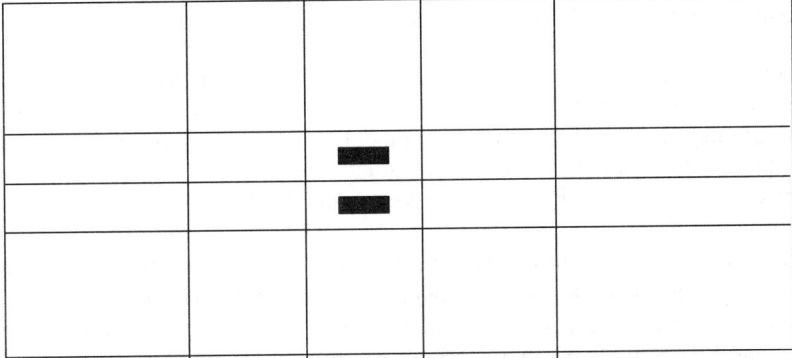

Figure 5:   Patching design of the catamaran's interface plane (seabed). Centre line parallel to x-axis. Six small innermost rectangles represent patches focal region, each rectangle 200m × 20m. This area is very finely meshed. Solid rectangles (roughly) represent catamaran's hulls. Diagram not set to scale.

on the interface plane alone necessary to satisfy our current internal point and element size conditions must therefore enumerate to more than 60,000. Now, this is not feasible with respect to either CPU run time or memory capacity, so an appropriate patching methodology must be derived.

The very fine meshing of the interface plane where the $d_{IP}:d_E$ ratio larger than of 0.25 is required within the reasonable vicinity of the hull which we shall now discuss. The length of L_{IP}, the line where the internal points are positioned, immediately reflects the size of the patches that are to be finely meshed. The patch assessment of the interface plane is schematically shown in fig. 5. The catamaran's hulls, if included in the picture, straddle $y = 0$ at the very centre of the plane. Ten narrow patches focus toward the centre of the plane, where $y = 0$. The six innermost and smallest of these rectangular patches are each 200m by 20m, and are subdivided by $360 - 3.33m \times 3.33m$ elements each. The $d_{IP}:d_E$ ratio is therefore taken as 0.3 within these patches. Also, even to continually ensure that we can obtain the good signature precision possible, *all* elements of the interface plane can remain of constant type. The four longer but equally slim patches are meshed 30×3, and the remaining 10 larger patches of the interface plane are meshed with 10×10 elements. Finally, the 'box' walls are also meshed relatively crudely by both 10×10 elements assigned to the rectangular patches below the level of the interface plane, and 1×10 elements for the narrow patches above the interface level.

When the meshing is entered to satisfaction, the final step is to dispense of nodes or mesh points that overlap with others, and this is done via the command MERGE_MP. This has the effect of leaving only one node at each original overlap location and effectively economising on the CPU run time as well as physical memory.

2.5 Introducing and identifying the zones

The model is now patched and meshed, and internal points have been accurately assigned wherefrom we will gather our electric signature results. The next step is associating regions, or zones, to our model, which are composed of a selection of patches maintaining a constant conductivity or resistivity. If we had no interface plane, then a single, default zone would be automatically assigned by the software.

In our model, we have the conductivities of water and the seabed to incorporate into the database. All patches on and above the interface plane will therefore represent the zone that will have the conductivity of water, accepted largely as 5.28 S.m^{-1} in Sydney Harbour. Similarly, the patches of the model on and below the interface plane will represent our seabed or the second zone, which has the conductivity of 0.528 S.m^{-1}. The patches of the first zone are all identified through 'outward' condition, which reflect the location of the (water) electrolyte. The seabed zone, in contrast with the water zone, must have the interface plane patches labelled as 'inward', indicating that the original or water electrolyte is on the other side of the interface plane. The remaining (pseudo-infinite 'box') patches are labelled 'outward', as these are correctly facing the material to which we intend to assign the seabed conductivity.

The zone location also determines whether symmetry will be assigned to specific planes. For example, Zone 1, the water zone, holds the xy (or $z = 0$) plane of symmetry, and the information that the model is said to be symmetric about $z = 0$ with respect to Zone 1. Although both zones claim a part of the zx- (or $y = 0$) plane, our full model has no symmetry in this direction. That is, if we only constructed the port side hull of the ship, then symmetry about $y = 0$ would be denoted in both Zone 1 and Zone 2.

It is also necessary to identify in which zone our internal points exist. Through our model, the internal points or our simulated sensors sit 1m above the seabed and are therefore in the sea zone or Zone 1.

Figures 6(a), (b) and (c) display a zoom-out perspective of the patch distribution of the entire model, Zone 1 and Zone 2 respectively.

2.6 Assigning boundary conditions and the material data file

All patches and hence elements must be assigned boundary conditions, with the exception of the interface surface. Boundary condition assignment will depend entirely on the materials that we wish to use and their status with respect to electric flux. We may hence generate an accurate location of the anodes, identify the condition of the steel hull, and denote the location of the NAB propeller.

The most convenient way to approach this step is to assign the prevailing hull material to every patch in the entire model. On assigning the boundary condition via patch, all elements on that patch will automatically share the same boundary condition. As already pointed out in an earlier subsection, the wall-patches in both zones of the pseudo-infinite box will have the condition of a zero normal current density. The propellers' patches are then altered from the steel

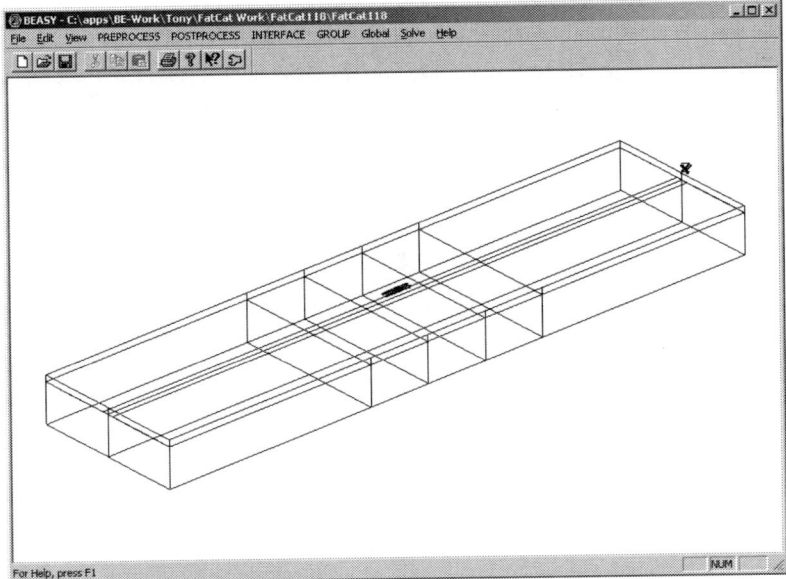

Figure 6(a): Patch distribution of model (esp. seabed & walls) before identifying zones.

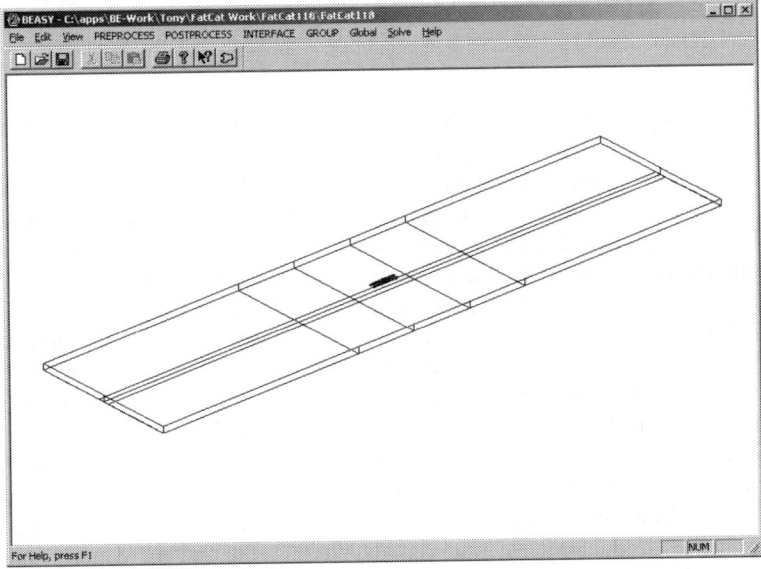

Figure 6(b): Zone 1 of catamaran model, composed entirely of outward-pointing patches of seabed, outer walls and catamaran itself (barely visible). Only symmetry of Zone 1 (and entire model) is about z = 0.

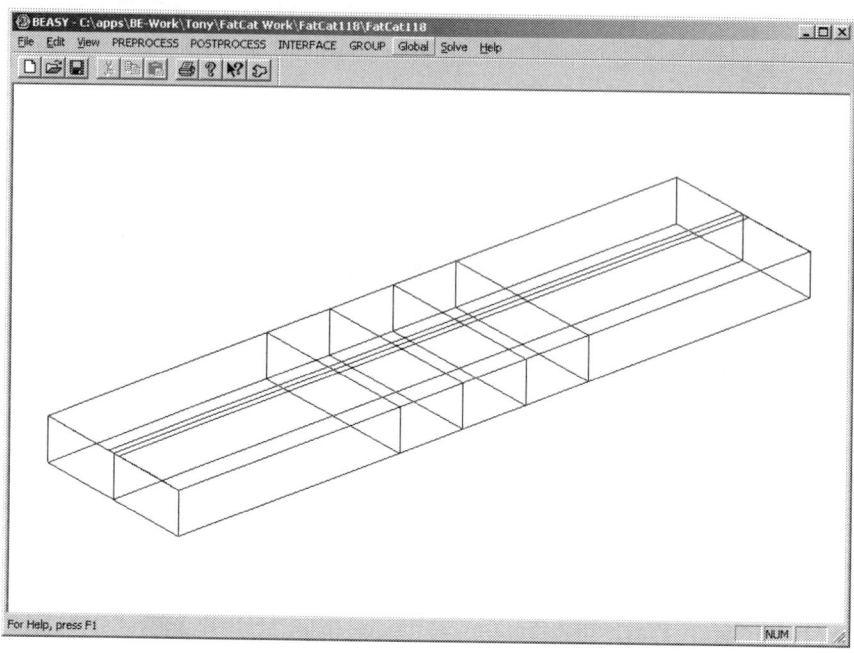

Figure 6(c): Zone 2 of catamaran model, composed of outward-pointing patches of outer walls and inward-pointing seabed. No symmetry defined in Zone 2.

originally assigned to NAB. The material boundary conditions described are through the BC_LABEL sub-command.

The focus is then on elements individually to allocate the flux conditions of the ICCP anodes. Although zinc anodes are not used in this model, the zinc is best defined by changing the materials of desired elements from steel to zinc. The tube element, also available in the BEASY code, is not as suitable to assign different materials to a hull surface as it will induce greater errors in the calculation. Figure 7 provides an illustration of the strip anode as replicated on the catamaran model. It shows that the anode surface is divided into 5 patches, and each patch itself is represented by a single element. The conducting strip, (usually) of platinised titanium is of 4m length and 2.54cm width, whilst the immediately surrounding region is PVC. Instead of simply replacing the materials of the elements, we must now represent the ICCP elements as a current source. The first step to take is to delete the existing boundary conditions of the anode patches and elements, and replace them with current density boundary conditions. If the current of one ICCP is –5Amps and the cross-sectional area of the conductor is $0.1016m^2$, then the current density, in $mA.m^{-2}$ will be set to –49,213. This is assigned to the element that represents the conducting strip, and the negative sign indicates that the current emanates away from the anode. The 4

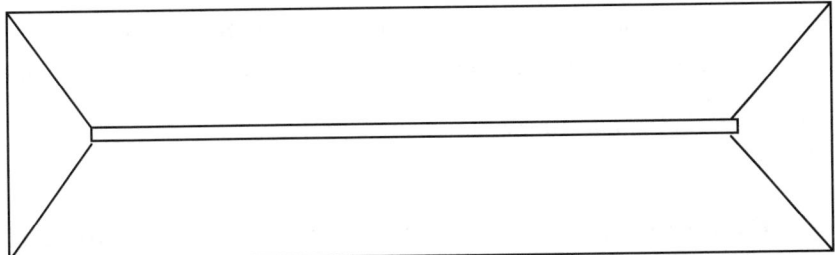

Figure 7:   Appearance of the ICCP strip anode as represented through BEASY modelling. The central strip represents the Platinised-Titanium conductor, 4m long and 2.54cm (1inch) in breadth, where J ≠ 0. Surrounding polygons are to be of PVC material with J = 0. Drawing not to scale.

surrounding elements are of PVC, and are therefore each assigned a current density of zero.

Polarisation data [5] associated with each material, the painted-steel hull and NAB, is identified to a ship via the BC_LABEL subcommand is a curve mapping a non-linear relationship between its electric potential against its electric flux density or normal current density. The polarisation data exists within the corresponding *.mat file of the model.

The BEASY model of the hypothetical catamaran is at this stage, almost ready for execution. The final step necessary is to essentially gather the data and include the parameters regarding the mode of execution. These parameters incorporate the tolerance, t, usually set to 0.01 and the desired number of iterations that is necessary for the maximum voltage error of the solution vector to fall below a number usually of the order of $3t$. These details create the *.dat file, of same name as the *.mat file, and the problem is therefore launched for solution.

2.7 The post-processor – extracting results

The post-processor is essentially the tool for extracting and interpreting the results of our boundary element computation. For our purposes, it is most appropriately adopted for the generation and display of the surface potential contour and the electric signature components, Ex, Ey and Ez. The graphical representation immediately viewed is the current density components, $\{Jx, Jy, Jz\}$ against the number of internal points along the line directly below keel, from bow to stern. The units of the current density components are mA.m^{-2}, and the conversion to μ V.m^{-1} simply applies Ohm's Law such that

$$E_n = \frac{J_n}{\sigma_w} \qquad (2)$$

where σ_w, the conductivity of the water in Sydney Harbour, is (already given) as 5.28S.m^{-1} or $5.28 (\text{Ohm.m})^{-1}$ and the subscript n denotes the vector component. Conserving dimensional balance, an electric field strength component is therefore 189.39Ohm.m times the current density of it's corresponding component.

3 ICCP currents, surface potentials and electric signatures of the catamaran

We shall now implement four ICCP strip anodes as described above on surfaces of the catamaran according to the array shown schematically in fig. 8. (The author once again wishes to point out that this is array, along with the model of the vessel itself, is purely hypothetical and any resemblance of this anode array to existing ones on similar vessels is purely coincidental). From fig. 8, the lower hull is referred to as the port hull, whilst the top one the starboard (stbd) hull.

Our objective is to protect the hulls from corrosion and at the same time observe the static electric signature induced through this process. We are limited to the use of 4 anodes and to the array shown on fig. 8 only. That is, we are not at liberty to shift the anodes in order to optimise CP and UEP signature reduction, only change the magnitudes of their currents. Realistically, the ICCP anodes of this model cannot be moved much further for or aft because the slim surface shape of the hulls cannot accommodate them. From our experiences with past warship boundary element models [2–4] greater emphasis must be placed on the stern regions due to the abundance of dissimilar metals (steel hull/NAB propeller), and hence very notable potential difference. The regions of both bows are essentially painted steel only and in theory, would not require a large anodic current to maintain an efficient CP and UEP signature reduction.

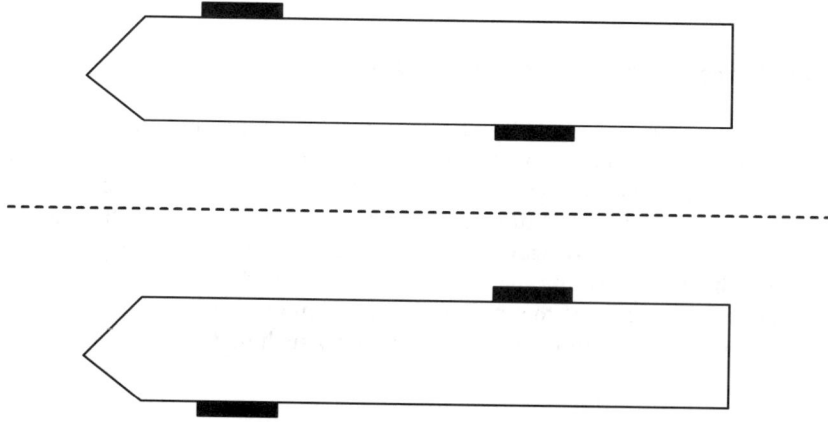

Figure 8:   Proposed ICCP anode array of catamaran. Broken centre line is segment of x-axis.

There are three sets ICCP current distributions used according to the given array. These are listed in table 1, for Settings A, B and C. For Setting A, the corresponding cathodic protection and UEP signature patterns are shown through figs 9(a) and 10(a) respectively, Setting B figs 9(b) and 10(b) respectively, and for Setting C figs 9(c) and 10(c) respectively. Figures 9(a) to 9(c) inclusive show the surface potential on Zone 1, that is, the hulls as well as the sea bed, depicted in the form of a colour contour. The scale to right is in mV, where the range is from –1,200 to –700mV. This scale is appropriate for Australian littoral waters, as a uniform hull potential that will provide a good cathodic protection is in the vicinity of –1,000 to –900mV [5].

Table 1: ICCP current settings.

Setting ID	2 For (Bow) Anodes	2 Aft (Stern) Anodes
A	–15Amps each	–15Amps each
B	–20Amps each	–20Amps each
C	–20Amps each	–30Amps each

3.1 Hull(s) surface potentials

The surface potential according to fig. 9(a) shows that the hulls only have cathodic protection close to the anode regions, implying that a magnitude of 15Amps/anode is insufficient to generate even a fair CP for larger surface areas on the catamaran hulls. The (red) stern region is entirely or well into the –700mV level, indicating a definite susceptibility to corrosion.

An interesting 'rule-of-thumb' feature can also be read from contours on the sea bed to predict the degree of continuity of the hull surface potential. In the case of fig. 9(a), the small tan ellipse beneath the bow might not only suggest that the total hull areas that share a potential between approximately –840mV and –800mV are quite small, but intermittent as well. In contrast, the much larger (red) contour below the stern could then imply that larger and more continuous regions at the stern of each hull will have a surface potential greater than –750mV.

With a need to increase the ICCP current magnitude to 20Amps/anode, fig. 9(b) displays the resulting surface potential contours. This shows a well-improved cathodic protection from bow to aft of midship hull regions, within the range –930mV to –880mV. From our rule-of-thumb, the huge (green) region on the sea bed confirms that a large area on the hull surfaces maintain the same range of potential, with of course, the exception of the anodes and the nearby regions. Moving further towards the stern regions, the uniformity, as well as the protection diminishes. The surface contour according to Setting B therefore illustrate that large hull regions are well-protected, though the stern regions still require a greater degree of protection.

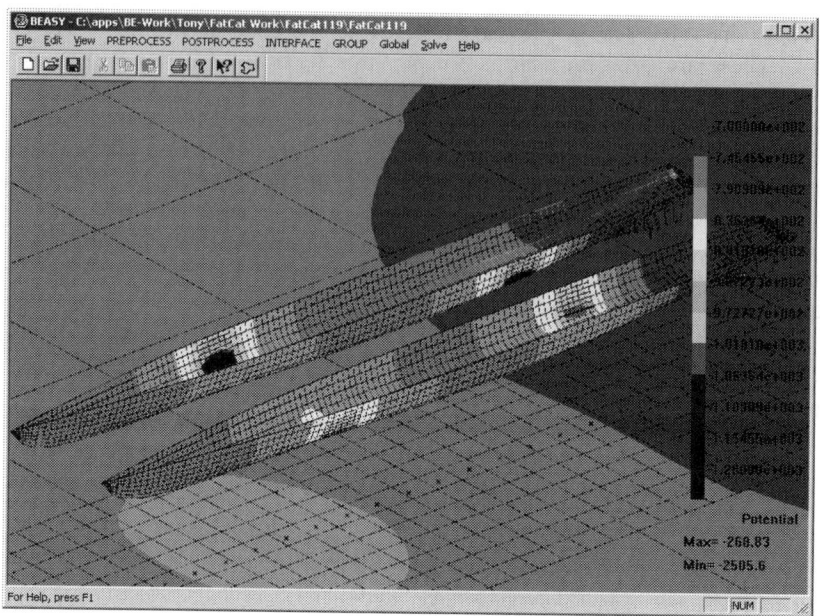

Figure 9(a): Colour contour of hull surface potential, Zone 1, when each anode
is set to −15Amps. Range is −1200 to −700mV.

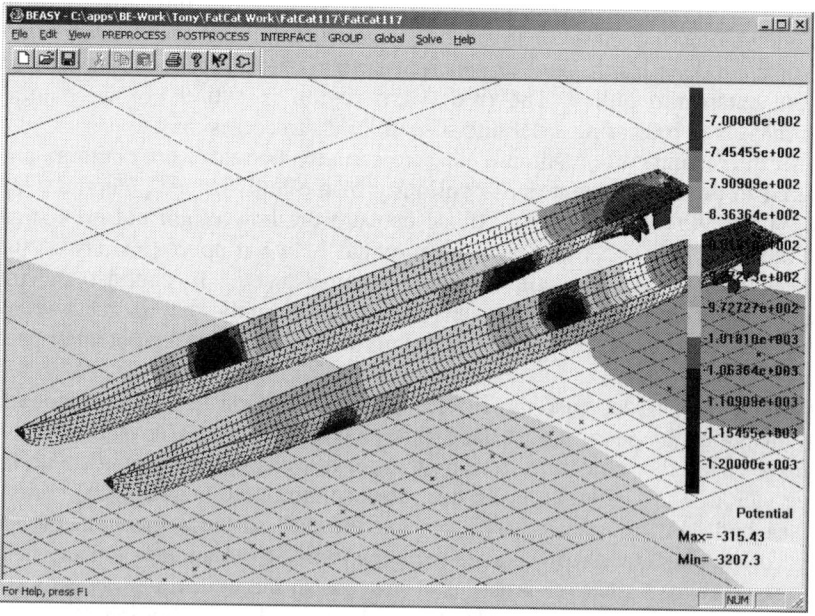

Figure 9(b): Colour contour of hull surface potential, Zone 1, when each anode
is set to −20Amps. Range is −1200 to −700mV.

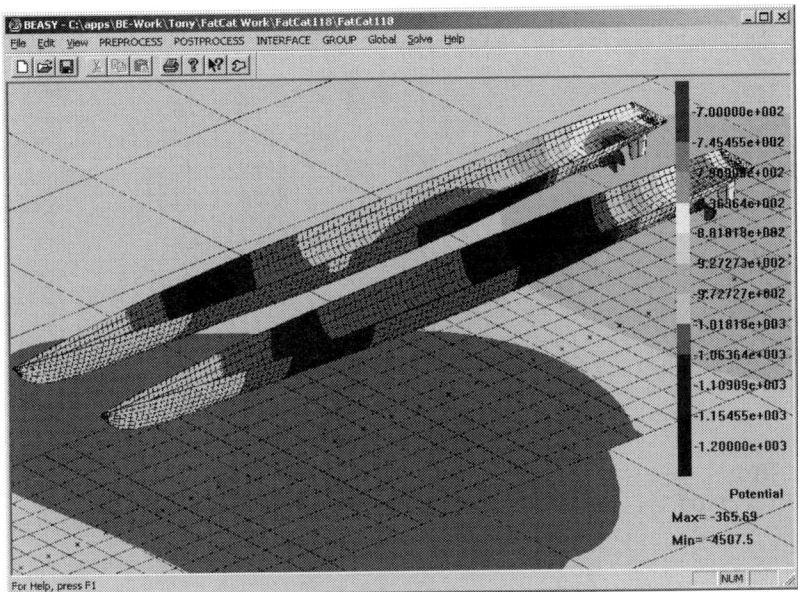

Figure 9(c): Colour contour of hull surface potential, Zone 1, when 2 aft (stern) anodes are set to –30Amps and 2 for (bow) anodes remain at –20Amps. Range is –1200 to –700mV.

The objective of Setting C is to prolong the excellent level of protection and its continuity whilst attempting to improve the degree of cathodic protection at the stern regions. This is why the current for anodes remain at 20Amps, whilst the current magnitudes of the aft ICCP anodes increase from 20Amps to 30Amps. Figure 9(c) shows that the majority of the hull surfaces (almost 90%) are protected, perhaps over-protected near the anodes. However, the range of the protected regions, (that is observing the blue and green regions), not including the anode regions themselves, is approximately –1,150 to –880mV. The larger the range or more appropriately, the larger the *difference* in the range, ΔV, the smaller the degree of continuity. This in turn has the effect of increasing the magnitude of the UEP signature, as discussed in the following sub-section. The median is approximately –1,020mV, and this is reflected through the moderately sized (blue) region on the sea bed. The positive result of Setting C is essentially the substantial decrease in the stern regions susceptible to corrosion.

3.2 UEP signatures

Figure 10(a) is the corresponding UEP signature where the internal points or 'sensors' are at a depth of 20m below the water surface and aligned for approximately twice the length of the hulls directly below the centre line or the *x*-axis. The solid, dashed and dotted lines correspond to the longitudinal, vertical

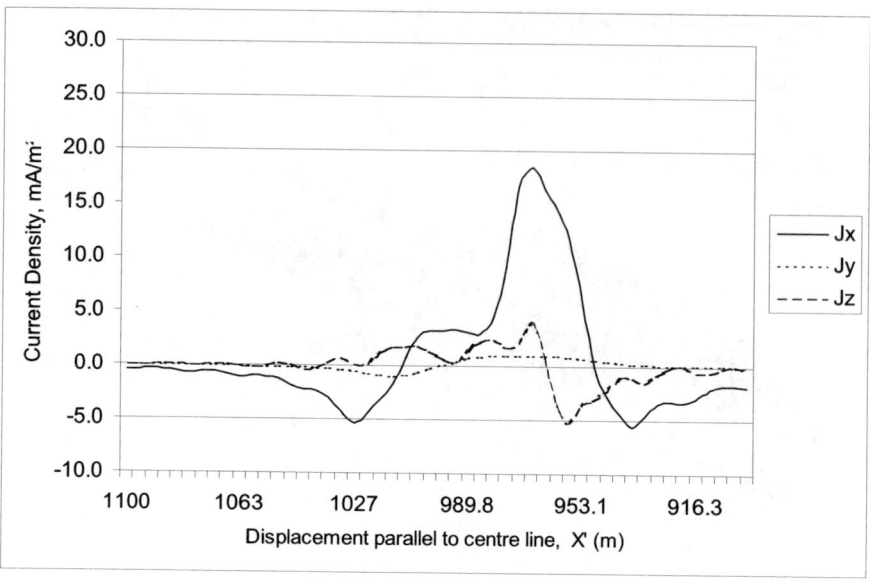

Figure 10(a): UEP signature when all anodes set to –15Amps.   'Sensors' along centre line or parallel to x-axis at depth of 20m.

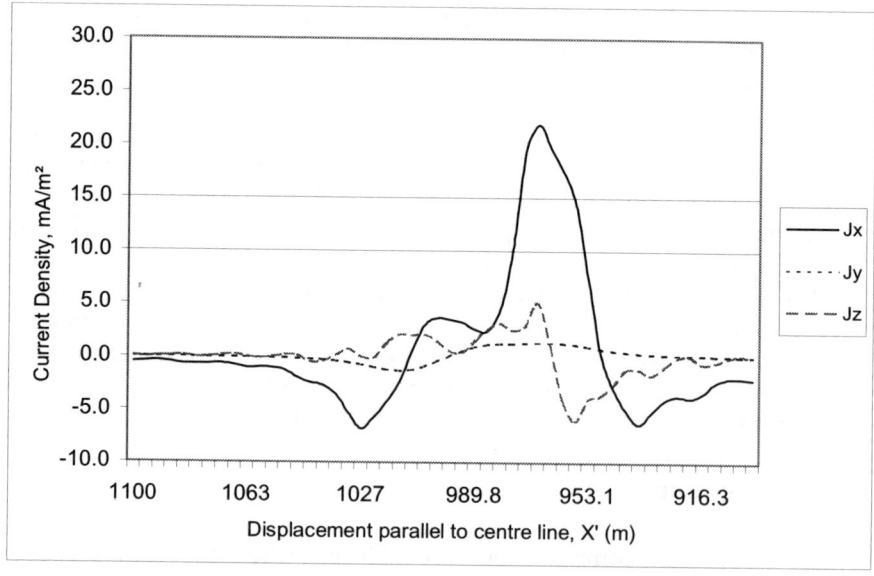

Figure 10(b):   UEP signature when all anodes set to –20Amps.   'Sensors' parallel to centre line or x-axis at depth of 20m.

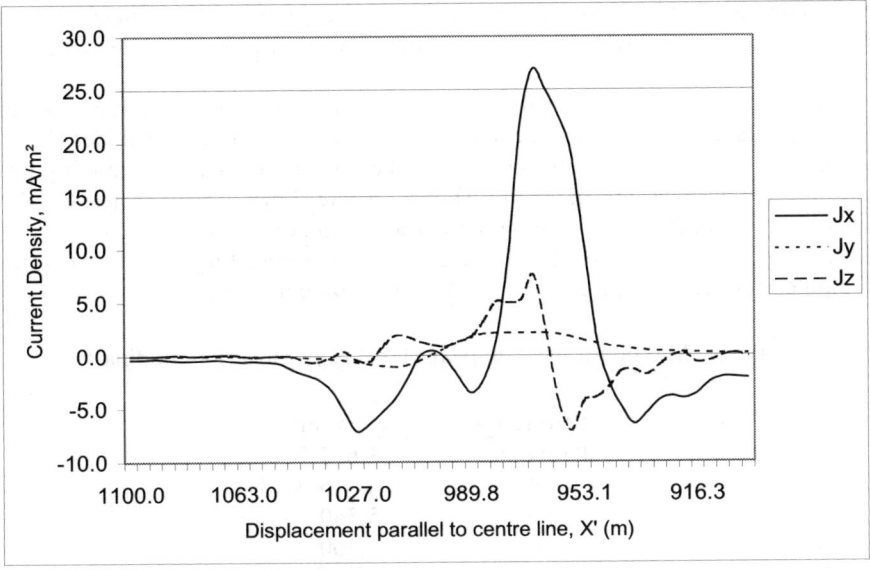

Figure 10(c): UEP signature when 2 aft (stern) anodes set to –30Amps and 2 for (bow) anodes remain at –20Amps. 'Sensors' parallel to centre line or x-axis at depth of 20m.

and athwartships components respectively of the current density as calculated at the given internal points. Each curve or UEP signature pattern commences about 50m in front of the catamaran's bows and ends around 50m behind the transoms. Figure 10(a) therefore depicts that Jx or Ex (from Ohm's Law above) is dominating component. The greatest disturbance or perturbation occurs in the stern region(s), where the curve peaks at approximately 18mA.m^{-2} ($\sim3,400\ \mu\text{ V.m}^{-1}$). The vertical component shows a very crude sinusoidal behaviour over the same domain, peaking at about 4mA.m^{-2} ($756\ \mu\text{ V.m}^{-1}$) and dipping close to -5mA.m^{-2}; ($-945\ \mu\text{ V.m}^{-1}$). The behaviour of the athwartships component is virtually negligible, and this is expected due to the mirror symmetry about the x-axis. Indeed, a forced symmetry would have given Jy and hence Ey as a straight line such that Jy (or Ey) is equal to zero throughout the entire domain. (Although symmetry would have also been favourable with respect to a much shorter run time and even better accuracy, it was purposely waived in this problem in order to observe any variations in the port or stbd hulls. This includes change of polarisation data to replicate random hull damage.

The increase in the ICCP current as per Setting B mainly demonstrates that there will also be an increase in the signature magnitude at the stern, especially recognised in the Jx curve of fig. 10(b). Figure 10(c), on illustrating the UEP signatures resulting from Setting C, principally shows that although an increase in the current magnitude may have a positive effect in reducing the magnitude of the stern regions highly susceptible to corrosion, it will also have the effect of increasing the signature peaks.

Figure 11 extracts each resulting *Jx* components for all three current settings and maps them on one graph for comparison. This graph also includes the *Jx* component when all ICCP anodes are switched off, (black curve), implying that the hulls themselves are of anodic status. Table 2 lists (peak to peak) UEP signature magnitudes for the longitudinal and component according to Current Settings A to C (inclusive). It also lists these resulting maximum when computed for zero-current ICCP's or an anodic hull condition.

Table 2: Maximum (peak to peak) UEP signatures (longitudinal).

Setting ID	Ex (uV/m)
0 (anodic)	1,657
A	4,376
B	5,340
C	6,300

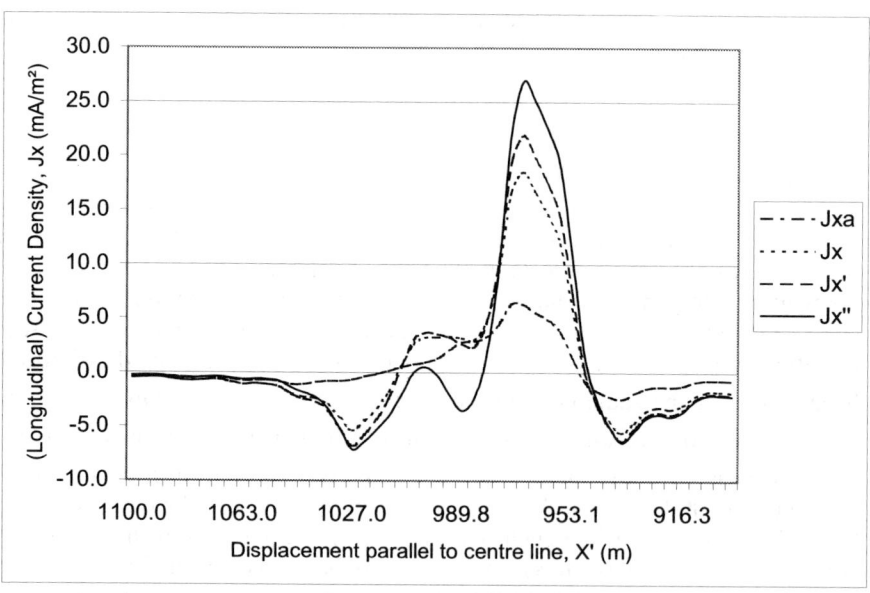

Figure 11:   Relative UEP signature magnitudes (longitudinal) at 3 anode settings: Jx (dots), all anodes –15Amps; Jx' (dashes), all anodes –20Amps and Jx" (solid) 2 aft anodes –30Amps & 2 for anodes –20Amps. The Jxa curve (dots-dashes) is the predicted UEP signature when all anodes are turned off *i.e.,* the hulls themselves are anodic.

4 Discussions and conclusions

We have investigated how we can accurately model and assess the cathodic protection and UEP signature issues of a multi-hull warship. The corresponding surface potential colour contours and signatures of our hypothetical catamaran have been calculated using BEASY according to several ICCP settings.

We have demonstrated that the best cathodic protection for the entire hull surfaces were obtained when the two fore anodes were set to 20Amps and the two aft anodes to 30Amps. Although the vast majority of the hulls received adequate protection, at least 5m^2 on the stern regions of each hull remained at risk to corrosion. In observing the hull profile as realistically as possible, we would anticipate moderately slim ones. This feature would therefore limit the available flat-surface hull area closer to the keel(s) and thereby also limit our freedom to redistribute a given number of ICCP (strip) anodes closer to the keel region. As we have seen, this is particularly important towards the stern regions, as it would be desirable to have anodes located relatively close to these areas that are at greatest risk to corrosion.

In the attempt to distribute currents to the stern regions, the high currents used have induced a very high UEP signature, which is the price paid for even a fair degree of cathodic protection in this region.

Possible resolutions for both the cathodic protection and signature problem are:

a) to retain the positions and currents of the two anodes, for anodes as per Setting C (or Setting B)

b) to replace the (aft) strip anodes with *two* disk anodes, and position the replacing pair closer to the risk areas, symmetrically about the keel. This will provide a total of three anodes per hull. The necessary current of the disk anodes can be adjusted through a trial and error approach. It has been shown in previous papers [2–4] that positioning anodes close to the high-risk regions will not only reduce the risk of corrosion but also reduce the signature magnitude.

Our interest would then be to observe maintenance, as the disk anodes would be subject to much greater turbulences regarding their position relative to the ship's propellers. This would prompt further research in fluid dynamics.

References

[1]   Trevelyan, J., *Boundary Elements for Engineers: Theory and Applications*, Computational Mechanics Publications, Boston, 1994.

[2]   Christopoulos, A.S., [Title RESTRICTED], UK CONFIDENTIAL – COMMERCIAL, Technical Report, DERA/SS/PSD(W)/TR000001/1.0, Winfrith, UK, 2001.

[3]   Christopoulos, A.S., [Title RESTRICTED], CONFIDENTIAL AGAO (AMRL Technical Report DSTO-TR-1210), Fishermans Bend, VIC, 2002.

[4]   Christopoulos, A.S., *Advancing Computational Prediction Techniques of Warship Corrosion Control and Electromagnetic Signatures*, Proceedings, Warship Cathodic Protection Symposium, Royal Military College of Science, Shrivenham, UK, 2001.

[5]   Millington, H.K., *Boundary Element Modelling of Warship Cathodic Protection Using BEASY – Achievements to Date with a Users' Guide* (U), UNCLASSIFIED, (PSL Technical Report DSTO-TR-XXXX), Fishermans Bend, VIC, 2002 (to be released).

Joint optimisation of corrosion protection and electric signature of a naval platform

Y. H. Pei
DSO National Laboratories, 20 Science Park Drive, Singapore 118230.

Abstract

The use of distinct materials on the outside of the ship hull produces electrochemical corrosion on the surfaces of these materials. In order to avoid severe corrosion damage at small isolated areas of paint, impressed current cathodic protection (ICCP) system is often used with painted surfaces. The use of corrosion protection systems generates electric currents flowing in the seawater which subsequently produce the electric and corrosion related magnetic signatures. The signatures may be used in the target classification of surface ships and submarines. The electric and corrosion related magnetic signatures might be significant if the ship's corrosion protection system is not designed and ranged optimally. To achieve an electric signature stealth naval platform, techniques such as computer simulation, design optimisation and physical scale modelling are the useful tools in solving the combined optimisation problem with complex ship structures. This paper presents a joint optimisation approach of corrosion protection and electric signature of a naval platform. We show that an electric signature stealth naval platform with sufficient corrosion protection may be achievable using optimisation technologies combining boundary element modelling and experimental approach. Inversion to the locations and strengths of the new corrosion areas when a ship is sailing in seawater is also proposed and inversion results are shown.

1 Introduction

Cathodic protection has been widely used for protecting structures from corrosion. The design of the cathodic protection systems, either sacrificial anode Cathodic Protection (CP) or Impressed Current Cathodic Protection (ICCP) system, require the solution of the Laplace's or Poisson's equation with the relevant boundary conditions to give the current and potential distribution in the solution domain. Even for relatively simple geometry, analytical solutions are usually not possible. Various numerical approaches have therefore been adopted and the Boundary Element Method (BEM) has been found to be the most proficient numerical method for the modelling of ship's cathodic protection system [12–16].

The design goal of an impressed current cathodic protection system is to produce an evenly distributed protection potential on the structures on the ship hull and major appendages. The available design variables are the number of anodes (and their location) and the location of the reference cells. The constraints on the design are the values of the potential on the structures. In order to provide adequate protection the corrosion potential must be less than a specified value (*e.g.*, –780mv). In order to prevent over protection the potential must be greater than a specified value (*e.g.*, –880mv).

The use of impressed current cathodic protection system produces the electric current flowing from the ICCP anodes to the corroded areas on the ship hull and major appendages and generates an electric field and corrosion related magnetic (CRM) signature [1–3]. To reduce the electric field and CRM signature, the optimum ICCP system design goal for a warship therefore is to achieve the requirement of producing a uniform distributed protection potential on the ship hull and to reduce the electric signature to a minimum.

Optimisation approach may be the solution to the joint requirements. The optimisation methods for electromagnetics have received considerable attentions in recent years. Many global optimisation methods, such as Genetic algorithm modelled on the principles and concepts of natural selection and evolution [7], Evolution strategies use the principles of simplified simulation of biological evolution as rules for the optimum seeking process [8], Simulated Annealing based on thermodynamic interpretation [9, 10], and the combinations of different global and determination optimisation strategies [7–10] were examined by combining the numerical computation procedures.

Computer simulations in ship corrosion prediction and ICCP system design have been studied [12–16]. Recent research efforts are focusing on the electric signature reduction using optimisation techniques combining computer simulations [4–6]. To achieve the design goals, BEM model can be coupled with global optimisation approaches such as Simulated Annealing (SA), Evolution, Genetic or other algorithms and deterministic approaches of both higher order and zeroth-order methods.

The complexity of a ship's corrosions and material polarisation makes it difficult to simulate real-life situations. In particular, the electric signature depends on the corrosions on the ship structures, dynamic conditions and operational environments during ship sailing. It is crucial that simulation models be verified with either the ship real measurements or the scale model experiments. During the last two decades, measurement technology has been developed in designing new and retrofitted ships [28–31]. The principle of the scale modelling is that the reduced linear dimensions of the scale model are compensated by an equal reduction in electrolyte concentration. Method of Dimension And Conductivity Scale Model (DACS) [28] is the most commonly used physical scale modelling technique.

2   The mathematical model

The governing equation in the bounded uniform seawater medium is [11]:

$$\sigma \nabla^2 u = 0 \tag{1}$$

where u is the electropotential, σ is the conductivity. The electric current density J is directly related to the first derivative of electric potential, u:

$$J = \sigma \nabla u \tag{2}$$

The boundary condition in the solution domain is defined by:

$$\Gamma = \Gamma_A + \Gamma_C + \Gamma_I \tag{3}$$

where Γ_A is the anodic surface, Γ_C is the cathodic surface and Γ_I is the insulated surface.

Commercial boundary element software BEASY-CP [17] was used in the numerical simulation.

The computation of corrosion related magnetic fields, the horizontal electric dipoles are immersed in the generalised three-layer seawater environment and the solution can be obtained by:

$$\nabla \times H = j + \sigma E \tag{4}$$

where j is the current sources. With definition of vector potential A in the solution domain, the electric field and corrosion related magnetic field is solved using:

$$\nabla^2 A - i\mu\sigma\omega = -j \tag{5}$$

The current source j can be defined as an electric dipole as the form:

$$j = I \delta(r_0) \tag{6}$$

where I is the strength of the dipole and r_0 is a position vector from the electric dipole.

Computer programs developed using analytical methods based on mirror image and integration [18, 19] for the generalised three-layers seawater environment was used in the computation of electric and CRM fields.

3   Computer simulation of polarisation

Polarisation is the relationship between electrical current and potential. The polarisation data describe not only the electrochemical reaction but also the environmental factors and can be generally expressed as [20–22]:

$$J = f(u, h, v, d, t, s, b, \text{etc})$$ (7)

where h is the film thickness, v is the flow velocity of the electrolyte, d is the depth, t is the temperature, s is the salinity and b is the bio-fouling. Other factors include oxygen content, etc.

Distinct materials may be used in the ship structure design. Ship hull, shafts and rudders may be made of steel, propellers may be made of nickel-aluminium-bronze alloy (NAB). Painting and coating on the steel surfaces is used to provide the initial avoidance from corrosion. With such designs, there is a significant risk of galvanic corrosion on the steel surface. As steel surfaces are coated, severe local attack can occur rapidly on any damaged surface due to the very high current density caused by the galvanic action.

To simulate the corrosion and the ICCP anode currents in operation environments, the computer simulation needs to consider both quiescent and dynamic material polarisation. For the slowly moving seawater, less available oxygen made its low concentration more important than the rates of oxygen reactions on the different metal surface. With the greater supply of oxygen reaction from seawater moving at the higher velocity the oxygen reduction reaction on the different metal surfaces has a major effect on the extent of galvanic corrosion. The current density to obtain a sufficient protection level depends on the seawater velocity and the degree of turbulence.

The hydrodynamic effects of the seawater flow on the surfaces of the structures are determined for different flow patterns [21–23]. Flow induced effect on the current density in cathodic polarisation of the steel in seawater, the following relationship was valid [15],

$$Sh = K\,Re^x\,Sc^y$$ (8)

where Sh is Sherwoud number, Re is the Reynolds number, Sc is the Schmidt number, $K = 0.07$–0.678 depending on type of flow, either laminar or turbulent, $x = 0.5$–0.68 depending on type of flow and $y = 0.33$ valid for different comparative flow patterns. By combining this equation with the equation for diffusion controlled O_2 reduction it was shown that the cathodic current density J_c could be expressed as a function of flowrate, all other parameters remaining constant [15],

$$J_c = C\,V^{\infty m}$$ (9)

where C is constant and $m = 0.5$–0.7 typically for surfaces without scale.

Material polarisation curves need to be experimentally determined for each material under given conditions. The accuracy of the BEM models is mainly dependant on the accuracy of the polarisation curves used. Simulation of real operation environments on metal samples to obtain representative polarisation data is difficult since the polarisation response of a material is affected by flow pattern, exposure time, physical scaling and interactions between the different materials.

In our computer model, the material polarisations [23] are used in the computer model and are modified by taking the influence of seawater velocity and temperature. The higher speed of seawater associated with turbulence flow conditions may increase significantly the rate of the electrochemical reaction compared with that observed for quiescent conditions.

The ambient temperature changes may increase or decrease the current. Literature data [20–23] were used in the computer models. The temperature and oxygen concentration of the seawater is taken as $30°C$ and $6\ O_2$ mg/l, respectively.

The computer model results which simulate dynamic condition using (9) were compared with measurement data [27] from a vessel using waterjet propulsion. The influences of seawater velocities on the wetted surface of the ship were considered in the computer model. The relative anode currents calculated as the anode currents during the ship travelling at different speeds to the reference anode current measured when ship speed at 25 knots is illustrated in fig. 1.

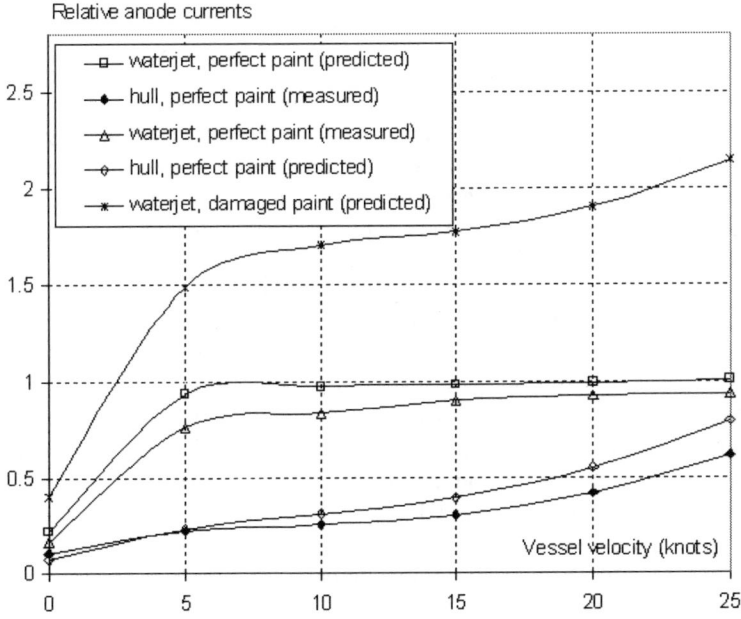

Figure 1:   Comparisons of simulated anode currents and measurement data.

4   Corrosion protection optimisation

4.1 Optimisation method

Consider a cathodic protection system with n degrees of freedom and an associated optimisation problem in the following standard form [24, 25]:

$$Minimise \qquad f(x) \qquad\qquad (10)$$

$$Subject \ to: \qquad \begin{cases} g(x) = 0 \\ h(x) < 0 \end{cases} \qquad\qquad (11)$$

where $x = (x_1, x_2, ..., x_n)$. The objective function $f(x)$: $R^n \to R$ is an optimisation criterion. The vector mappings g: $R^n \to R^r$ and h: $R^n \to R^s$ define the constraint equations and constraint inequalities, respectively.

The objective function and constraints in the application are defined as:

$$Minimise \qquad f(i, s, u) = \sum_{n=1}^{N} (u_n - u_t)^2 \qquad\qquad (12)$$

$$Subject \ to: \qquad \begin{cases} u_{min} \le u_n \le u_{max} & n = 1, 2, ..., N \\ v_{min} \le v_m \le v_{max} & m = 1, 2, ..., p \end{cases} \qquad\qquad (13)$$

where i is the anode electric currents, u is the potentials at the measurement points, s is the locations of the anodes, N is the measurement points of corrosion protection potentials on the ship hull, u_{min} and u_{max} are the minimum and maximum protection potential level on the protection surface and are set up as –780 and –880mV *Ag/AgCl* respectively, $u_n = (u_1, u_2, ..., u_N)$ are the potentials at the measurement points which are required to satisfy the constraint $u_{min} \le u_n \le u_{max}$, u_t is the threshold potential and is defined as –830mV *Ag/AgCl* on the reference cell, $v_m = (v_1, v_2, ..., v_p)$ are the required measurement points on the corrosion protection surface by a optimisation process, usually $p \ge N$.

The objective function is a measurement to the uniformity of the potential distributions on the corrosion surfaces to be protected with a particular anode arrangement, corrosion condition and operational environments.

The tasks of optimisation for ICCP design are:

- Select the number of anodes.
- Guide the positioning adjustments of the anodes through an optimisation process.
- Adjust the anode currents by an optimisation process to fulfil the corrosion protection on the ship hull and appendages.

In order to validate the suitability and effectiveness of the optimisation methodology, both stochastic and determination optimisation methods were investigated:

- The stochastic optimisation: Simulated Annealing (SA).
- The determination optimisation: higher order and zeroth-order deterministic methods.

Deterministic optimisation methods [24, 25] are the standard tool for solving local optimisation problems of simpler structure. Unfortunately, most real optimisation problems are global due to the complicated structure of the solution domain and highly non-linear phenomena. The stochastic optimisation approach proposed in this study is "Simulated Annealing" which is a computer simulation method based on a strong analogy between the physical annealing process of solids and the problem of solving large combinatorial optimisation problem [26]. The algorithm tries to find the global optimum of an N dimensional function. It moves both up and downhill and focuses on the most promising area. Starting from an initial point, it takes search steps along every co-ordinate direction and a starting "temperature". The "temperature" is a control parameter of the SA algorithm.

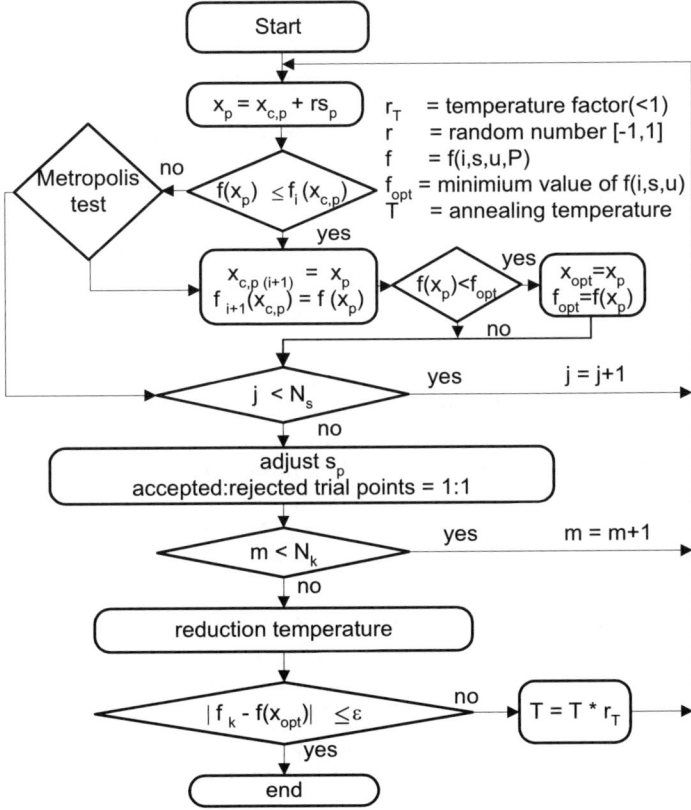

Figure 2: The flow chart of the proposed SA algorithm.

The main advantage of this approach is the substantial independence from the specific design problem. The only inputs to the optimisation method are a proper cost (objective) function whose minimisation leads to a solution of the problem, and simple "tuning" of the SA parameters in order to improve the search according to the function behaviour. The main disadvantage is the high computation cost due to the large number of functional evaluations. The flow chart of the proposed SA algorithm is shown in fig. 2. However, considering the complicated structure of the solution domain, specially the evaluation requirement for the objective functions of a multi-zone ICCP system, it is necessary to combine the global optimisation with the deterministic optimisation to accelerate the optimisation process. In the optimisation process, the deterministic method may be used to achieve the actual minimum while the SA is used to identify the promising areas of global minimum.

4.2 Corrosion protection optimisation of a sample ship

To verify the effectiveness of proposed optimisation method, a sample warship with a length of 35m was investigated. The geometry of interest is the wetted surface of the hull and major appendages. ICCP system evaluated includes two pair of anodes and two centre controlled power supply. In the present design, the ship has two propellers and rudders. The propellers are made of NAB alloy and modelled as solid disks with equivalent surface area as the real elements. The rudders and shafts are made of steel. The propellers, rudders and shafts are assumed to be uncoated because of turbulence engendered by propeller movement. The ship hull is also made of steel which is coated to provide initial protection from corrosion.

The definition of the paint status, locations of damaged areas of paint in the computer model are based on observations of the same and similar class of ships in dry-dock after the ships have served a certain period of time. Paint and coating are normally damaged on parts of the hull, especially on the aft zone around propellers. The observed painting damages on the hull, specially on the surface closing propellers are that the corroded areas have many smaller bare areas and corroded points. The equivalent conductivity of the damaged paint surfaces is therefore based on the ratio of the bare steel to that of the painted surface. The equivalent conductivity of the damaged paint surface was then modelled as a percentage of the polarisation response of steel. The perfect painted steel hull was modelled as either insulated surfaces or very small percentage of steel polarisation.

The mesh discretisation was designed to represent the most likely positions that the anode will be placed. The ship is surrounded by infinitive seawater and was modelled by a boundary of the solution domain far away from the region of interest. The seawater was defined with a constant conductivity of 5 S/m in the computer model. Accuracy of the electrochemical analysis is dependent upon the material polarisation involved in the boundary element model. In this computer model, linear representation of the polarisation which matches the service conditions was used.

The objective of the SA algorithm is to evaluate the global minimum and distinguish the local minima. In this investigation, the anode position is defined based on the computation result from SA algorithm. The anode location is then moved based on the SA result and the boundary element mesh discretisation is kept unchanged. In this case, the SA algorithm was used to evaluate the objective function $f(i,s,u)$ with the constraints. The minimum distance between two vicinity anode locations in this examination is set up roughly as 1.0m apart and this found to be the good representative to the global map of the investigated optimisation problem.

The normalised objective function map (fig. 3) shows the effect of anodes located at different positions on the surface of the sample ship while there is only one pair of anode in the SA evaluation. To minimise the computational cost in the SA process, the boundary element model for ship hull is divided into sub-regions in which the potential is assumed to have the same value. Each SA movement and boundary element computation result is recorded. The recorded data is reused in the objective function evaluation if the SA algorithm movement belonged to the same sub-region. Using this approach, the repeat effort of boundary element computation is avoided.

The SA algorithm was run over 4 times with initial coordinate $x = (20, 3)$, $(8, 2)$, $(5, 3)$, $(15, 1)$ respectively. The evaluation surface is on the wetted ship hull and $x = (x_1, x_2)$, x_1 counts from of the stern towards the bow and x_2 counts from water line towards the keel line of the ship. The global minimum was found in each case and SA algorithm terminated successfully on the global minimum irrespective of where the initial anode was located. The SA movements are about 200 for each case and the effective SA movements related the initial coordinates are 29, 25, 18 and 34, respectively.

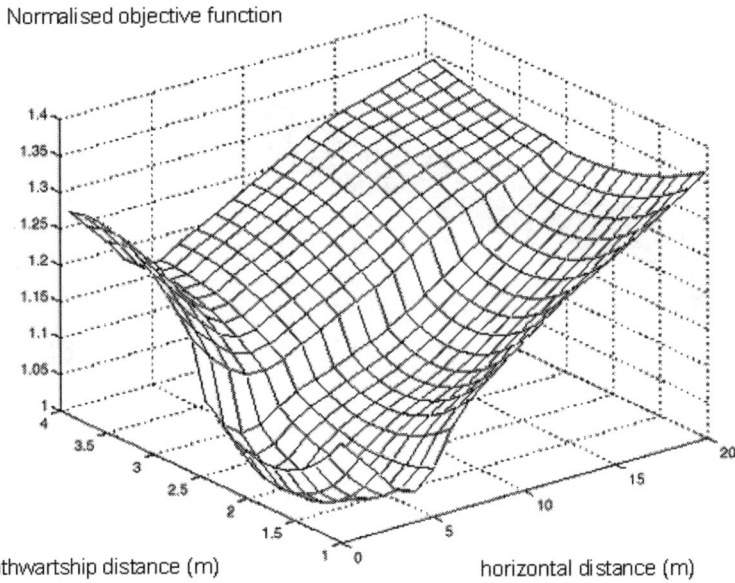

Figure 3:   Normalised objective function map on ship's wetted surface.

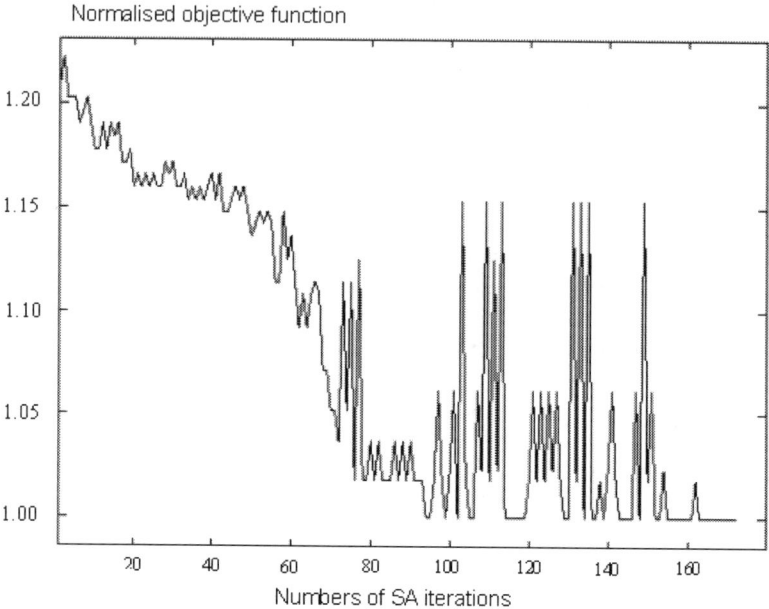

Figure 4:   SA process of convergence of normalised objective function.

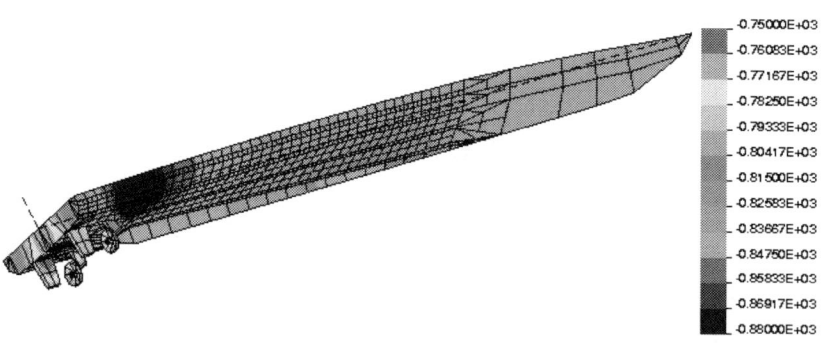

Figure 5:   Potential distribution on the ship hull with anode location of
$x = (3, 3)$.

The process of convergence of the normalised objective function when the initial coordinate is located at $x = (15,1)$ is illustrated in fig. 4. The SA algorithm moves to the promising global region and searches between local and global minima. The global minimum is found successfully where the constraints are being obeyed even if the location between the local and global minimum is very close. The objective function is shaken in the vicinity of the global minimum location since the complexity characteristics of the geometric on the aft zone. The potential distribution on the wetted hull surface with the optimum anode location is illustrated in fig. 5. The improvements of the corrosion protection potentials may be observed through the observation of the corrosion protection potentials along ship's keel line before and after optimisation as illustrated in fig. 6.

Having obtained the patterns of corrosion potentials on the ship hull and the sensitivities of anode locations and currents from the studies of BEM modelling and global optimisations, pre-determining a region which contains the desired objective function minimum is achievable and thus determination method can be implemented. To validate the deterministic methods, two pair of anode system was considered. One pair of anode on the front portion of the hull is allowed to move around the front area and another pair anode on the aft portion is allowed to move around the aft area. The adjustments of the position of the anodes are based on the sub-regions where the potential of each sub-region is assumed to be constant. The discretisations of the sub-region may be determined based on the mesh structure and the investigation for gradients of potentials on the ship hull.

The higher order deterministic methods which use the gradient of the objective function need less objective function evaluations and are therefore faster. However, the higher order deterministic algorithms were found to be unstable due to the complicated objective patterns and the convergence is dependent upon the starting points. We found that the downhill simplex and one-dimension search methods may need more objective function evaluations and therefore are less efficient. However, it is more robust in the implementation of this complicated optimisation problem and may be more suitable.

A comparison of the optimisation process using different deterministic methods and starting points is illustrated in fig. 7. With different starting points, deterministic methods may trap into local minima, where the one-dimensional search either reached the global minimum or trapped at the local minimum. Comparison of potential along keel line of different anode locations is illustrated in fig. 8. The better corrosion potentials on the ship hull is achieved by this optimisation approach. The optimum anode locations and corrosion protection potentials on the ship surface is illustrated in fig. 9.

5   Joint optimisation of corrosion protection and electric signature

To reduce the electric field and CRM signatures, the optimum ICCP system design goal for a warship is to achieve the requirement of producing a uniform distributed protection potential on the ship hull and to reduce the electric signature to a minimum.

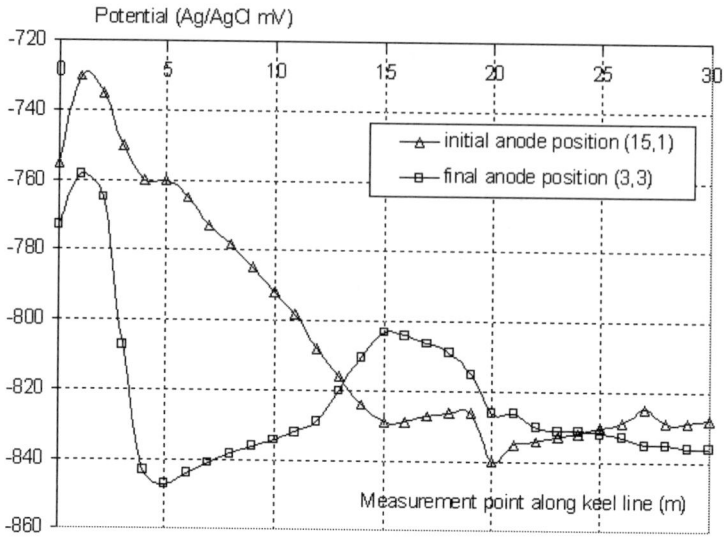

Figure 6:   Potential on the ship's keel line before and after optimisation.

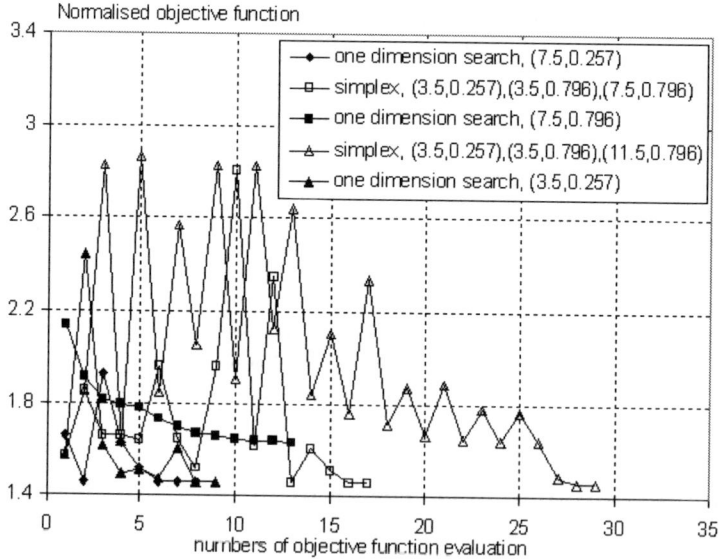

Figure 7:   Comparisons of convergence of one dimension search and down hill simplex with different starting points.

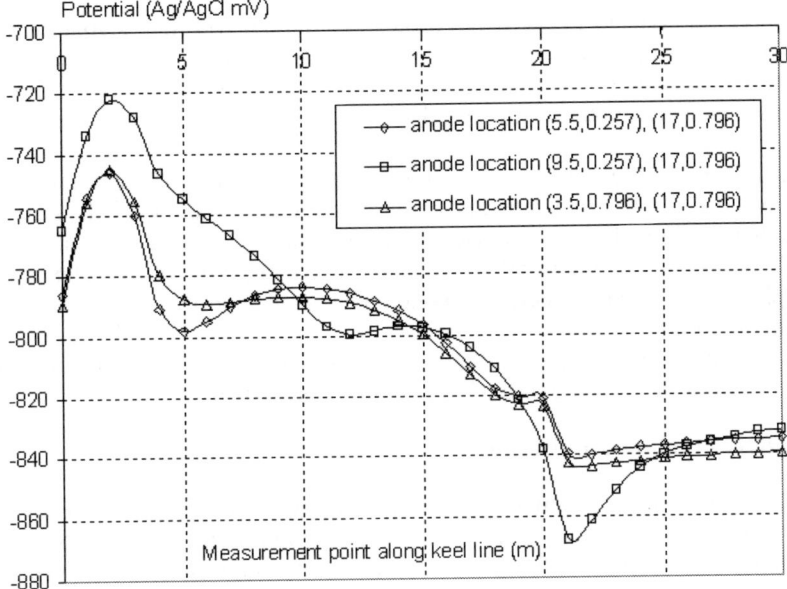

Figure 8: Comparison of potential along keel line of different anode locations.

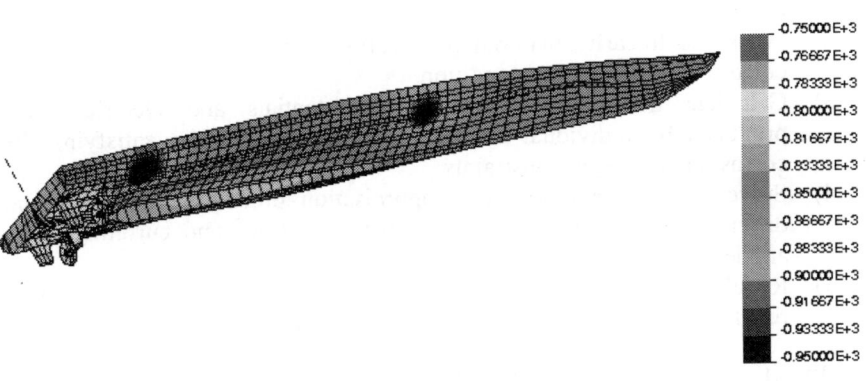

Figure 9: Final anode locations and corrosion protection potentials on the ship hull.

The joint requirement to corrosion protection and electric field signature may be difficult to a ship which only corrosion protection requirement is considered and only a smaller number of anodes are available. In such situation, only limited electric signature reduction ratio could be achievable by optimising the anode currents.

To achieve a better electric signature reduction with sufficient corrosion protection, the number and location of the anodes and reference cells may be needed being defined and optimised in design phase. The electric signature minimisation could be then realised through a ranging process by determining the amplitudes of the anode currents based on a specified corrosion condition and seawater environment.

Both optimisation design and ranging process of a low electric signature warship could be achievable through a joint optimisation approach.

5.1 Method of joint optimisation

The joint optimisation is achievable through an optimisation process combining with BEM models. The problems facing for the optimisation is the extensive computation time used in BEM modelling due to the non-linear of the material polarisation curves. To increase the speed of computations in optimisation process, one possible way is to assume that the linearity material polarisation curves could be utilised at the operation points based on a specified corrosion condition and operation environment.

With this hypothesis, the procedures of the joint optimisation may be as the following:

1) Pre-determine the premising positions of the anodes in reducing the electric signature with an adequate corrosion protection. The determination of anode positions could be achieved through BEM computation or physical scale modelling, or combining the two approaches.
2) Work out linearity material polarisation curves based on a specified corrosion and operation condition.
3) Calculate the corrosion protection potentials and electric fields produced by individual anode using BEM models by satisfying the corrosion protection constraints.
4) Solve and compare the joint optimisation objective functions from different combinations of the numbers, positions and currents of the anodes.
5) Redefine a specified corrosion condition based on operation requirement. Repeat the joint optimisation process from 2) to 4).

5.2 Objective functions and constraints

We may define the objective function as to minimise the electric fields measured on a surface below the ship bottom and to maintain the sufficient corrosion protection potentials on the ship hull and appendages.

The primary objective function is to achieve a sufficient corrosion protection.

Under linear assumption of the solution system, the potentials produced by individual anode on the measurement points on the ship hull may be written in a matrix form of,

$$
\begin{bmatrix} u_1 \\ \vdots \\ u_n \\ \vdots \\ u_N \end{bmatrix} = \begin{bmatrix} u_{1,1} & \cdots & u_{1,m} & \cdots & u_{1,M} \\ \vdots & & & & \\ u_{n,1} & \cdots & u_{n,m} & \cdots & u_{n,M} \\ \vdots & & & & \\ u_{N,1} & \cdots & u_{N,m} & \cdots & u_{N,M} \end{bmatrix} \begin{bmatrix} i_1 \\ \vdots \\ i_m \\ \vdots \\ i_M \end{bmatrix}
\tag{14}
$$

where N and M are the total number of reference cells and anodes, respectively, u_n is the corrosion protection potential on the reference cell, i_m is the anode current, $u_{n,m}$ is the unit corrosion protection potentials on the reference cells produced by m^{th} anode current.

The second objective function is to minimise the electric field signature. Under linear assumption of the solution system, the electric fields produced by individual anode may be written in a matrix form of,

$$
\begin{bmatrix} e_1 \\ \vdots \\ e_l \\ \vdots \\ e_L \end{bmatrix} = \begin{bmatrix} e_{1,1} & \cdots & e_{1,m} & \cdots & e_{1,M} \\ \vdots & & & & \\ e_{l,1} & \cdots & e_{l,m} & \cdots & e_{l,M} \\ \vdots & & & & \\ e_{L,1} & \cdots & e_{L,m} & \cdots & e_{L,M} \end{bmatrix} \begin{bmatrix} i_1 \\ \vdots \\ i_m \\ \vdots \\ i_M \end{bmatrix}
\tag{15}
$$

where L is the total number of measurement points of electric fields on a surface below ship bottom, e_l is the electric field on the measurement point, $e_{l,m}$ is the unit electric fields on a surface below ship bottom produced by m^{th} anode current and $\vec{e} = \left(e_x \vec{i} + e_y \vec{j} + e_z \vec{k} \right)$ is 3-axis electric field vectors.

By accumulating the potentials and electric fields produced by each anode current, the objective functions can be constructed as,

$$
Minimise \qquad f_u(i,s,u) = \sum_{n=1}^{N} \left(\sum_{m=1}^{M} u_{n,m} \, \hat{i}_m - u_n^t \right)^2
\tag{16}
$$

$$
Minimise \qquad \max_{l=1,2,\cdots,L} \; f_e(i,s,u) = \sum_{m=1}^{M} e_{l,m} \, \hat{i}_m
\tag{17}
$$

where u_n^t is the measured potential at the reference cells, $\hat{i}_m$ denotes the estimator of anode current i_m.

In order to utilise an optimisation algorithm with multi-objective functions (16) and (17), the objective functions are summarised to an integrated function with a penalty parameter α as,

$$\textit{Minimise} \qquad f_{eu} = f_e + \alpha f_u \qquad\qquad (18)$$

α to be determined through testing the balance of objective functions and the convergence of the optimisation solution.

The objective function are subjected to the following physical constraints,

$$\textit{Subject to:} \quad \begin{cases} u_{\min} \leq u_n^t \leq u_{\max} & n = 1,\ 2,\ \cdots,\ N \\[2mm] \hat{i}_m \geq 0 & m = 1,\ 2,\ \cdots,\ M \\[2mm] \bar{u}_t = \displaystyle\sum_{n=1}^{N} \frac{u_n^t}{N} \end{cases} \qquad (19)$$

where $\bar{u}_t$ is the average value of measured potentials at reference cells and is set up as Ag/AgCl -830mV. Since the ICCP system has an automatic feedback control to the corrosion protection potentials based on the measured potentials at reference cells, an adaptive iteration has to be adopted in the optimisation process to modify the anode currents to satisfy the corrosion protection in (19).

With above definitions for the objective functions (18) and constraints (19), the problem under consideration is to solve an inverse problem so as to calculate approximately the corrosion potentials and the electric fields of each anode under a particular corrosion condition. The minimisation of the objective function under constraints is to find the optimum combinations of number, positions and currents of anodes.

In general, the inverse problem under consideration may be ill-posed and there is no guarantee of the uniqueness of the solution. To obtain a stable numerical solution of the inverse problem, statistical formulation or regularisation methods could be used in order to reduce the uncertainty of the inverse solution.

5.3 Optimisation results

The sample ship in the optimisation is a 1.05m mock-up ship. The conductivity of seawater is set up as 0.05 S/m based on Dimension And Conductivity Scaling (DACS) system. Following DACS scaling principle, we simulate a ship of 105m at conductivity of seawater 5 S/m.

The sketch diagram of BEM model of the mock-up ship is illustrated in fig. 10. Detailed structures of the mock-up ship including propellers, rudders are included in the BEM model.

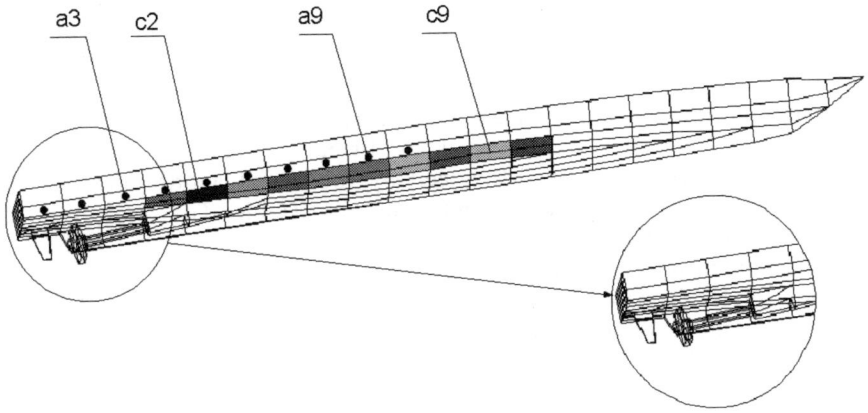

Figure 10: BEM model of the mock-up ship.

Based on preliminary BEM simulations, the anode positions in the BEM model were determined. There a total of 10 anodes are arranged at equal intervals and located toward the stern of the ship and labelled as a1 to a10. The 10 corrosion areas are defined in the BEM model and are labelled as c1 to c10. The corrosion areas are located symmetrically on the two sides of the hull. The area of each corrosion surface is about 0.05m x 0.025m. The corrosion on these areas can be defined as bare surface or partial corroded surface.

Several cases were investigated to verify the features of the optimisation approach. Two cases demonstrated are:

- Case 1: investigate the perfect coating, only propellers and shafts are treated as bare surfaces.
- Case 2: investigate the damaged coating, propellers, shafts, rudders and corrosion areas c5, c6 and c7 on the hull surface are treated as bare surfaces.

5.3.1 Case 1: Joint optimisation for perfect coating

In this investigation, only propellers and shafts are modelled as the bare surfaces. This scenario is to demonstrate a ship with the new paint status.

In solving the optimisation problem, we need to minimise electric fields on a surface below ship hull and achieve evenly distributed corrosion protection potentials on the ship hull surface subject to the constraints. Global optimisation approach SA was selected to solve the optimisation problem with multi-objective functions and constraints.

In the first step of the optimisation, corrosion potentials on the hull and electric fields on a surface below the ship bottom produced by individual anode were calculated by BEM modelling. Vertical directional electric fields produced by each anode below ship bottom 300mm along keel line are illustrated in fig. 11.

The SA iteration processes with setting equal initial anode currents are illustrated in fig. 12. The objective function in iteration process is displayed as per hundred iterations an interval in the plot. The SA needs about 10^4 to 10^5 iterations to achieve a stabilised solution and the computation time is about 1 to 2 minutes in a Pentium IV PC. The different initial anode currents were tested as listed in table 1. Though the convergence speed may be influenced by the initial anode currents, the optimum anode currents were found to be very stable. The final optimum anode currents are listed in table 2.

The optimum anode currents and objective function in each SA optimum iteration (or local minima of anode currents) are illustrated in fig. 13. In the computation process, the SA algorithm tests every possible combination of the anodes and anode currents and compares the objective function obtained up to latest iteration. The solution is the objective function to achieve the minimum and to be stabilised.

The electric fields are reduced to a large ratio comparing with the initial anode currents as illustrated in fig. 14. The electric fields are illustrated 300mm below the ship bottom along the keel line.

Corrosion protection potentials on the surfaces of the ship hull of initial anode currents and optimum anode currents are compared in fig. 15. A better corrosion protection may be achieved using the optimisation approach.

The reduction of the electric fields may need the considerations not only the anode current distributions but also the number of the anodes since the number of anodes has to be minimised. Therefore the tasks of the optimisation are to optimise the anode currents and also select the number and the position of the anodes.

To study the selections of the anode number and position, several cases were examined.

The optimum anode currents are found to be not equally distributed among the 10 anodes and large amount currents are concentrated in 2 anodes closing to corrosion areas. Optimum electric fields using only 4 anodes closing to stern was investigated. The optimum anode currents are listed in table 3 and the SA optimum process is illustrated in fig. 16.

The optimisation was examined in the case if only 2 anode available in the ICCP system. This is the case of existing ICCP system that uses one anode at aft position and another one at front position of the ship hull to cover the corrosion protection. The aft anode is located at anode a2 and the front anode is located at anode a10.

The influences of the electric fields of different anode currents are investigated as listed in table 4. For the equal anode currents between anode a2 and anode a10, the electric fields is significantly larger than the case of only aft position anode switching on, since the anode a2 closes the cathodic cells and the shorter electric current path minimises the electric moment.

The SA optimum iteration process for 3 different initial anode currents is illustrated in fig. 17. The electric fields changes against anode currents between the 2 anodes may be observed also in these curves.

Table 1: Initial and optimum objective function.

Anode No.	Initial anode currents (%)			
	1	2	3	4
1	10.46	0.00	0.00	0.00
2	10.71	52.28	104.55	0.00
3	10.66	0.00	0.00	0.00
4	10.39	0.00	0.00	0.00
5	10.18	0.00	0.00	0.00
6	10.05	0.00	0.00	0.00
7	9.96	0.00	0.00	0.00
8	9.91	0.00	0.00	0.00
9	9.88	0.00	0.00	0.00
10	9.87	53.55	0.00	104.55
Initial objective function	1.23	1.27	0.37	2.22
Optimum objective function	0.34	0.34	0.34	0.34

Table 2: Optimum anode currents.

Optimum anode currents (%)									
1	2	3	4	5	6	7	8	9	10
0.0	87.45	19.55	0.0	0.0	0.0	0.0	0.0	0.0	0.0

Table 3: Optimum anode currents for 4 anodes.

Optimum anode currents (%)				
Anode No.	1	2	3	4
Anode currents	0.00	87.45	19.55	0.00

Table 4: Initial and optimum objective function (anode a2 and a10 switch on).

No.	Initial anode currents (%)		Optimum anode currents (%)		Initial SA Function	Optimum SA function
	Anode a2	Anode a10	Anode a2	Anode a10		
1	52.276	53.551	104.550	0.000	1.271	0.371
2	35.412	70.825	104.550	0.000	1.595	0.371
3	0.000	104.550	104.550	0.000	2.221	0.371
4	104.550	0.000	104.550	0.000	0.371	0.371

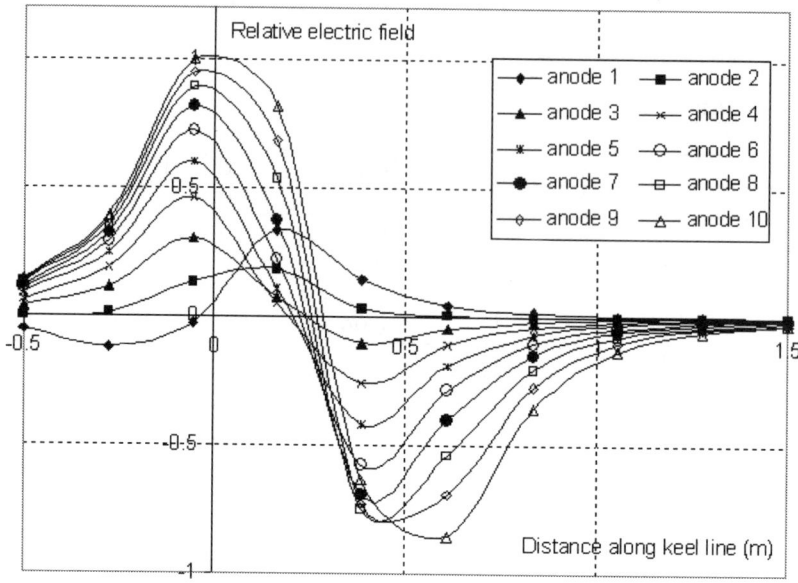

Figure 11: Vertical electric fields produced by each anode.

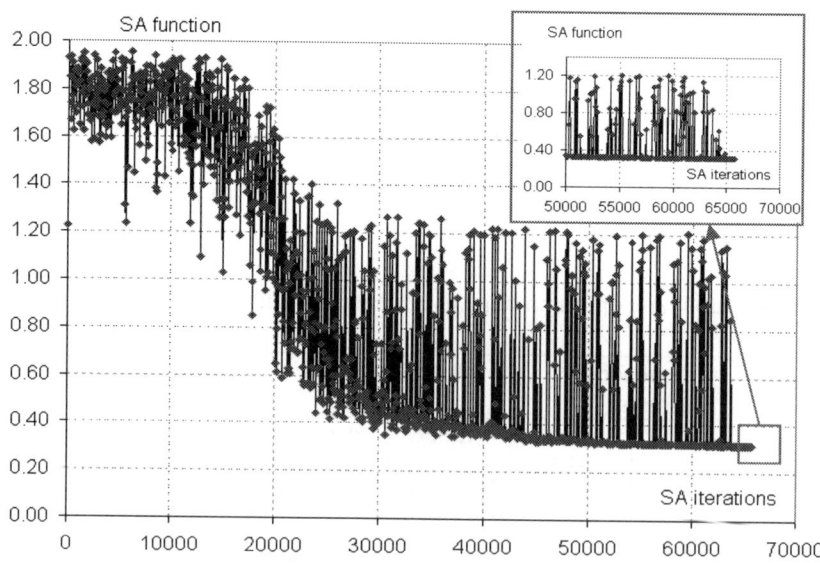

Figure 12: SA iteration processes.

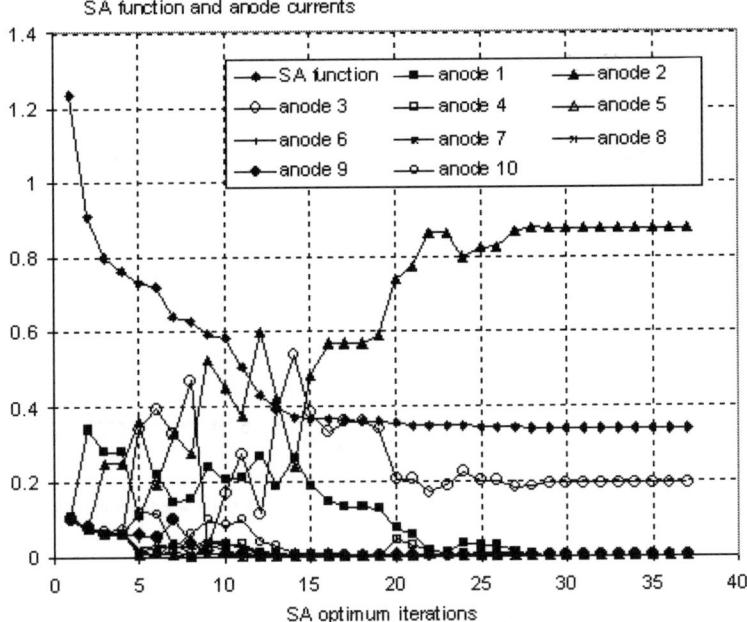

Figure 13: SA iteration processes for optimum functions.

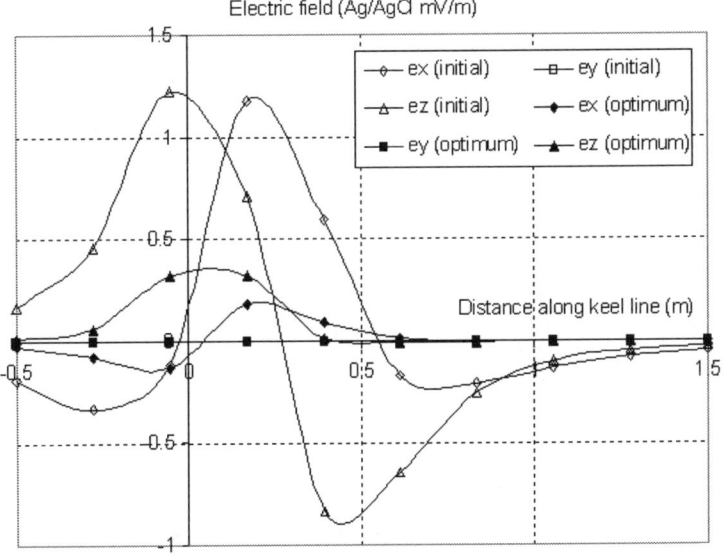

Figure 14:   Electric fields before (equal initial anode currents as in case 1 of table 1) and after optimisation.

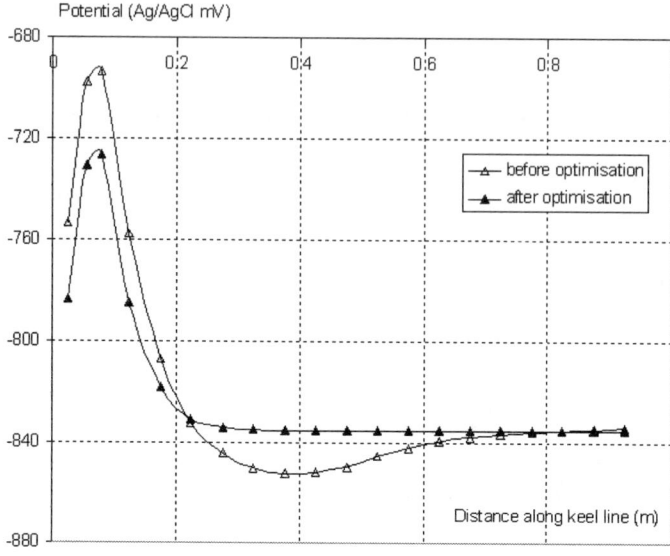

Figure 15: Corrosion protection potential on the hull along keel line before (equal initial anode currents) and after optimisation.

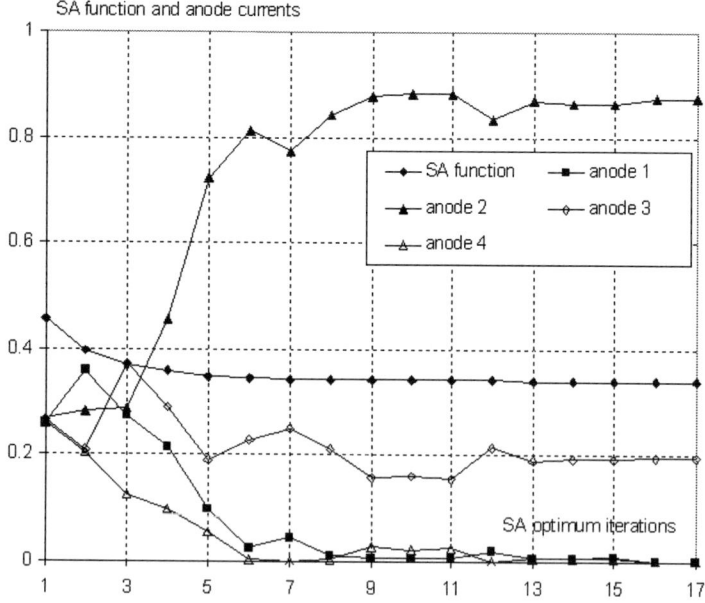

Figure 16: SA iteration process using only 4 anodes for optimum functions.

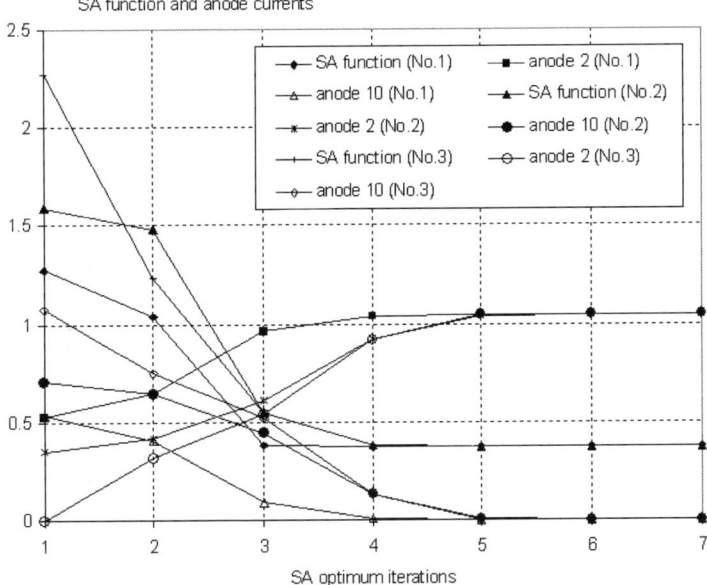

Figure 17: SA optimum iteration process, only anode a2 and a10 switching on, corresponding to table 4.

5.3.2 Case 2: Joint optimisation for damaged coating

In this investigation, corrosions for the damaged coating were examined. The propeller, shafts, rudders and corrosion areas c5, c6 and c7 on the ship hull are modelled as corroded areas in the BEM model. This scenario is to represent the old paint for a warship having sailed a longer time in the sea.

The electric fields on a surface below the ship bottom and corrosion protection potentials on the ship hull are calculated by BEM modelling. The vertical electric field produced by the individual anode along keel line below ship bottom 300mm is illustrated in fig. 18.

The achieved objective functions against different initial anode currents are listed in table 5. In the case of only anodes a2 and a10 are switched on with equal anode current, the objective function is 0.918. In the case of only anode a2 is switched on the objective function is 1.614 and in the case of only anode a10 is switched on the objective function is 2.426. The results show that positions of the anodes and distributions of the anode currents are sensitive to the electric fields.

Final optimised anode currents are listed in table 6. The results of anode currents show that the anode currents are concentrated at the corrosion areas, one group of currents is concentrated in aft portion and another group is concentrated around c5, c6 and c7 corrosion areas.

The SA iteration process is illustrated in fig. 19 and SA optimum iteration process is illustrated in fig. 20. Similar SA iterations and optimum iteration processes were found compared with perfect paint case.

Table 5: Initial and optimum objective functions.

Anode No.	Initial anode currents (%)			
	1	2	3	4
1	10.986	0.000	0.000	0.000
2	11.109	54.930	0.000	109.860
3	11.116	0.000	0.000	0.000
4	10.790	0.000	0.000	0.000
5	10.484	0.000	0.000	0.000
6	10.362	0.000	0.000	0.000
7	10.431	0.000	0.000	0.000
8	10.575	0.000	0.000	0.000
9	10.632	0.000	0.000	0.000
10	10.580	55.546	109.861	0.000
Initial objective function	0.734	0.918	2.426	1.614
Optimum objective function	0.326	0.326	0.326	0.326

Table 6: Optimum anode currents.

Anode No.	1	2	3	4	5
Optimum anode currents (%)	15.11	36.31	1.46	5.72	2.04
Anode No.	6	7	8	9	10
Optimum anode currents (%)	0.13	7.68	18.95	20.52	0.31

The 3-axis electric fields before and after optimisation is illustrated in fig. 21. The electric fields before optimisation are under the condition of equally initial currents and the electric fields are measured at 300mm below the ship bottom along the keel line. The final optimisation results show that the electric signatures are improved significantly through the optimisation approach even with a complicated corrosion condition. The effects of optimisation may be more significant comparing a traditional ICCP system with 2 anodes with equal anode currents.

Corrosion protection potentials on the ship hull surfaces of initial anode currents and optimum anode currents are compared in fig. 22. An even corrosion protection may be achieved using the optimisation approach.

The tasks of the optimisation are to optimise the anode currents under a specified number of anodes and anode positions. This is to be a time consuming optimisation process since the choices are missive. The optimisation depends on the anode numbers, positions and the corrosion conditions of the ship.

Selected assessments to the selections of numbers, positions of the anodes are listed in table 7. The objective function may be slightly worse than 10 anode cases if only 5 anodes are switched on.

A typical case is illustrated in fig. 23 and fig. 24 where only anode a2, a6 and a10 switching on. The results show that the electric fields can be reduced 58.7% by optimising the anode currents.

Table 7: Selected anode numbers and optimum anode currents.

Case No.	1	2	3	4	5	6
Anode No.	Anode currents (%)					
1	30.69					
2		60.21	58.87	53.71	66.37	109.86
3	28.16					
4		0.00				
5	0.00		0.00			
6		0.00		37.92		
7	21.40					
8		42.40	51.59			
9	28.19					
10		6.18		18.87	43.98	
Initial objective function	0.53	0.98	0.62	0.94	0.92	1.61
Final objective function	0.33	0.33	0.37	0.39	0.65	1.61

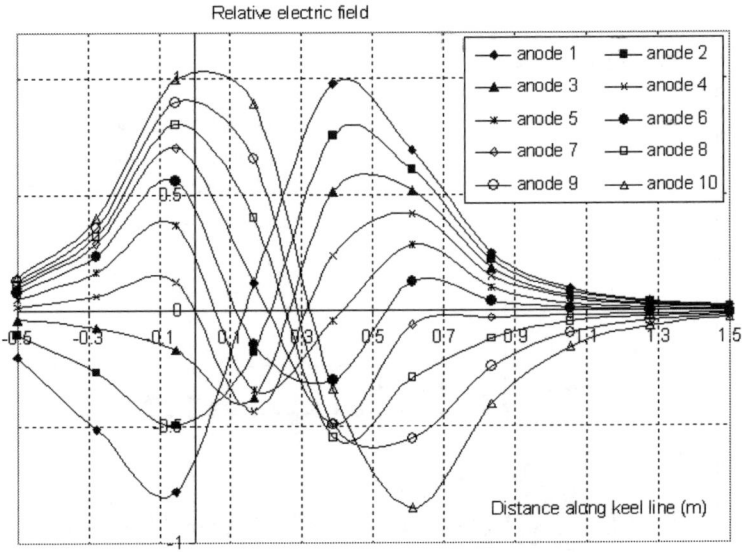

Figure 18: Vertical electric field below water line.

6 Physical scale modelling

Scale modelling has long been used as an accurate and cost-effective method of simulating the characteristics of prototype physical systems in the many fields of sciences. Despite the advances made in numerical modelling techniques, the

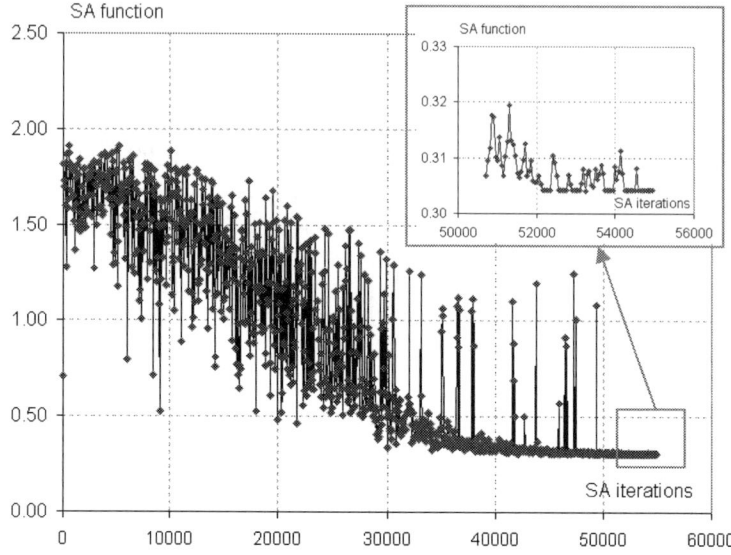

Figure 19: SA iteration processes.

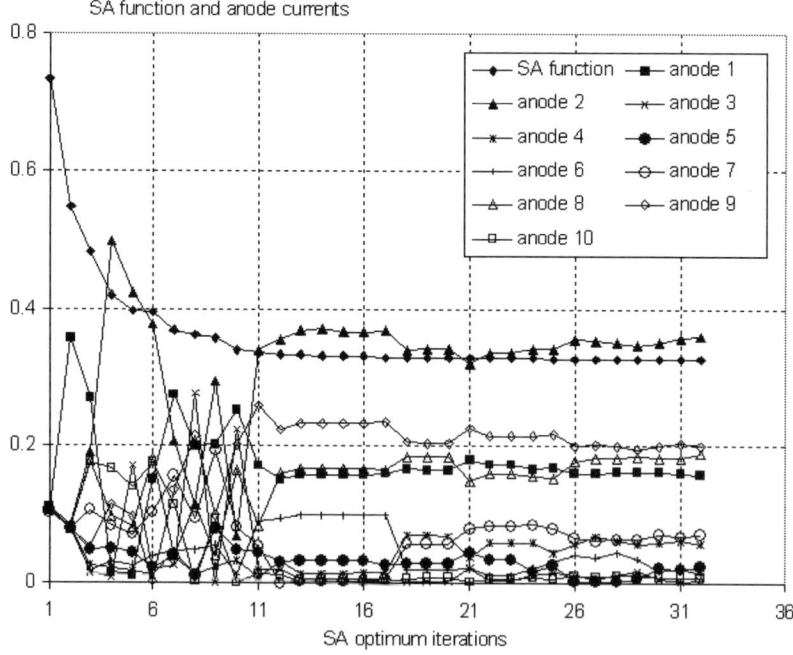

Figure 20: SA iteration processes for optimum functions.

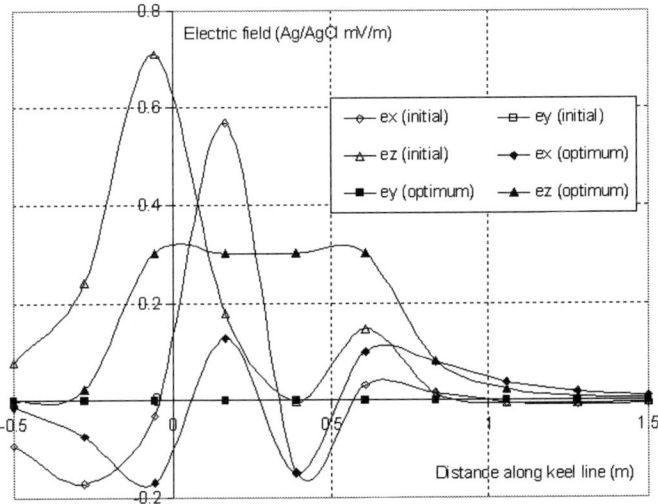

Figure 21: Electric fields before (equal initial anode currents) and after optimisation.

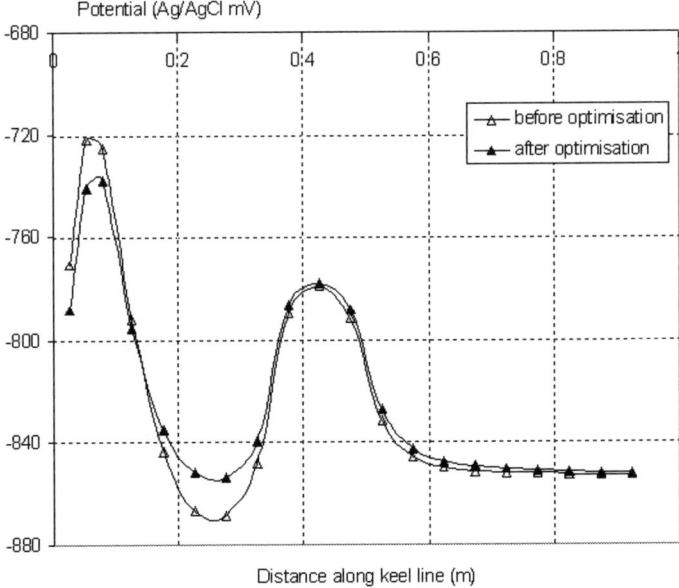

Figure 22: Corrosion protection potential on the hull along keel line before (equal initial anode currents) and after optimisation.

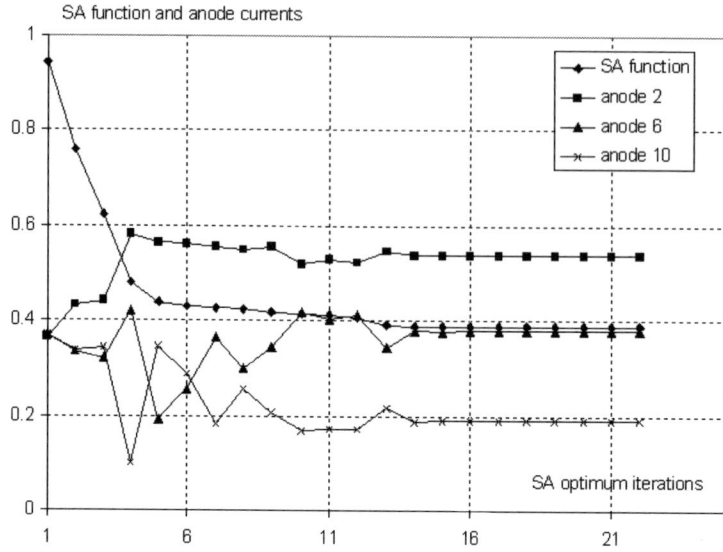

Figure 23: SA iteration processes for only anode a2, a6 and a10 switching on.

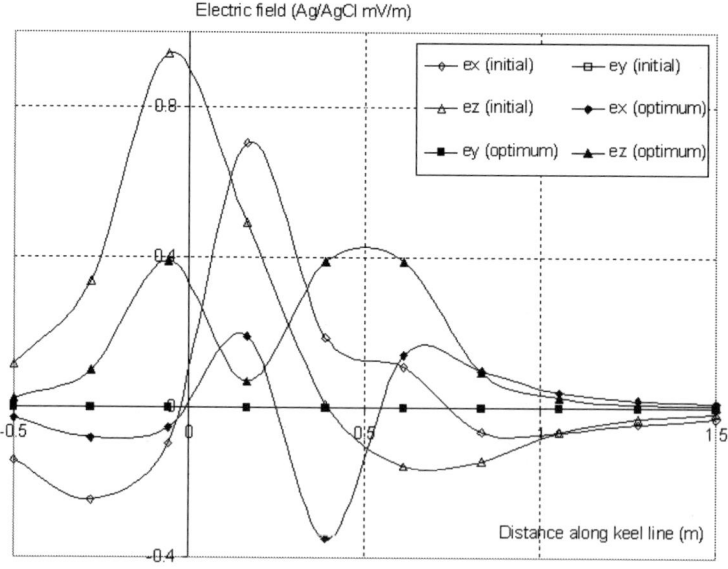

Figure 24: Electric fields (anode a2, a6, a10 switching on) before (equal initial anode currents) and after optimisation.

experimental validation of theoretical predictions 'taking into account the very complicated geometry and uncertain boundary conditions' is still an important step. The use of physical scale modelling to incorporate the effects of fluid flow on the surface electrochemical potentials of cathodically protected structures, together with the associated underwater electrical potentials, requires a precise understanding of the relationships between the scale model characteristics and those of the real structure. The proper conducting of a scale measurement requires careful consideration of several scaling laws in order to ensure that model measurements reflect the true operational conditions. In general, Dimension And Conductivity Scaling (DACS) [28] are most common techniques in the physical modelling to the ICCP system of ship.

In this experimental technique, metal scale models constructed of the same materials as the prototype are floated on or immersed in seawater which has been diluted, such that its electrical conductivity has been reduced by the same factor as that used to scale the linear dimensions of the model. Typically, one-sixtieth to one-hundredth scales are employed. Scale anodes are attached, with the DC power provided by a power supply and under way conditions are simulated by flowing electrolytes. Surface potentials over the structure are measured by an array of silver/silver chloride (SSC) electrodes attached to the model. Such approaches to scale modelling have been validated against warship data taken during hull potential surveys. The technique has also been used to examine the underwater electric fields around protected structures by employing a remote pair of SSC electrodes to measure potential differences in the electrolyte. Results can be presented as potential difference profiles at various distances from the structure or as maps of the magnitude and direction of the electric field.

The DSO DACS platform is illustrated in fig. 25. The DACS platform is a fibreglass water tank, 3m in length, 2m in width and 1.8m in height. To simulate the appropriate hydrodynamic conditions, the hydrodynamic simulation system includes a fibreglass water channel of 215mm in radius and a water control system with pipes, valves and a velocity meter. Two pumps each capable of 3.7kW output provide the primary drive and the maximum water speed is 3 m/s in this channel.

In the measurement process, the electric sensor can be moved along a rail track of straight lines with a controllable speed and the positions of the sensor motion can be recorded in a marker system. A 3-axis electric sensor is used in the measurement of electric fields. The electric sensor is made of *Ag/AgCl* and the separation of two potential probes is 40mm. During the measurement, the measured electric potential and electric field signatures are filtered through a 14 channel analogue filter and data are collected through a data acquisition system and computer. The measurement process is illustrated in fig. 26 and the measurement equipment is illustrated in fig. 27.

A 1.05m mock-up ship is scaled down to 1/60 to 1/100 of the real ship. The structures used in this study are showen in fig. 28. The ICCP structures have been determined by BEM modelling combined with the optimisation approach. The structure has nine corrosion areas (c1–c9), which are located symmetrically on the two sides of the hull except c1 on the bottom of the hull close to the propellers. The area of each corrosion surface is 0.05m x 0.025m and can be opened

Figure 25: The measurement of anode currents with simulated hydrodynamic
condition in DSO DACS platform.

Figure 26: Measurement of potential and electric field of mock-up ship in
process.

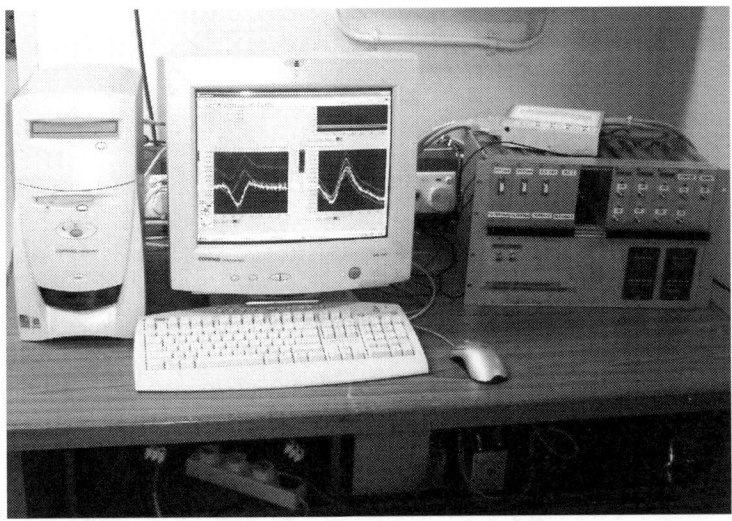

Figure 27: Measurements equipment in DSO DACS.

during the measurement. The corrosion on these areas can be conditioned as partial or completed corroded surface. There are a total of 18 anodes arranged symmetrical on the two sides of the hull (a1–a9). 4 anodes (a1–a4) are located on the stern close to the propellers with smaller interval (50mm) and the other 5 anodes are located with large interval (100mm) toward the bow of the scale ship. A total of 8 Ag/AgCl reference cells (r1–r8) are attached on the keel line of the scale ship to measure the surface potentials. Reference cell c1 is located on the hull close to the propeller and the other 7 reference cells are located along the keel line from the stern toward the bow with the same interval (100mm).

The scale ship has 2 braze propellers driven by 2 motors and the rotational speeds can be controlled from 5 to 280 rpm. Two steel rudders are installed and these rudders can be rotated to simulate the hydrodynamic resistances.

Corrosion protection potentials and electric fields have been measured to the mock-up ship with different corrosion, anode currents and hydrodynamic conditions.

The mock-up ship has been immersed in the scaled conductivity seawater, with no pre-exposure to full strength seawater, for over 1 week before any measurements. In the measurement process, reference cell r2, which is located at the edge of the keel line near the shaft, is selected as the reference corrosion protection potential and the reference potential is set up as *Ag/AgCl* –830mV.

Two typical measurement results are presented.

6.1 Measurement 1: Static corrosion condition

In this measurement, static corrosion condition was assumed and measured with static scale seawater environment. The corrosion surfaces on the mock-up ship are the propellers, rudders, shafts and the corrosion surfaces c7 and c8 which are located on the front part of the ship hull. The anode currents were

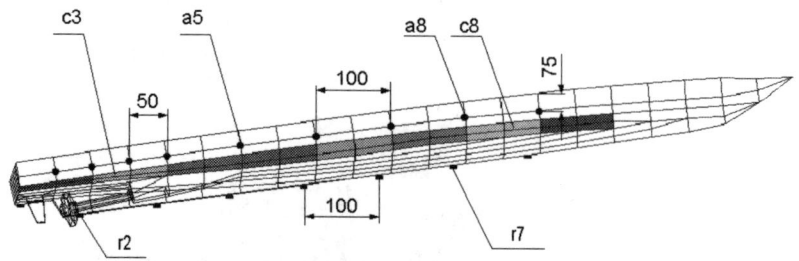

Figure 28: Structures of scale ship used in DACS measurements.

adjusted to keep the potential at reference cell r2 to *Ag/AgCl* –830mV. The electric fields were measured accordingly through the electric sensor moving along the longitudinal axis below the mock-up ship. The processes of minimising electric fields were trial and error by observing the changes of electric fields during each adjustment to the anode currents. The trial and error processes were also guided by the experiences obtained from the simulation studies of joint optimisation approaches using BEM and optimisation technique.

The joint optimisation processes in the DACS scale model are illustrated in fig. 29. The comparisons of electric fields before and after optimisation are illustrated in fig. 30, where the initial electric fields are on the condition of switching on anode a5 and a9 with equal anode current. It is observed that the electric fields can be reduced to a large ratio by the optimisation process compared with the initial setting of the anode currents.

6.2 Measurement 2: dynamic corrosion condition

In this measurement, the dynamic corrosion condition was assumed and simulated through moving water around the mock-up ship at a speed about 2.0 m/s. The corrosion surfaces are the same with measurement 1, but the propellers are rotating at 280 rpm. The joint optimisation processes are illustrated in fig. 31. The comparisons of electric fields are illustrated in fig. 32, with the same condition of measurement 1. Both anode currents and the electric fields increase the dynamic condition compared with the static condition. The electric fields can be reduced to a large ratio in the dynamic corrosion condition through the optimisation process, though the locations and amounts of anode currents may be different from the static corrosion condition. The phenomena observed in the DACS scale model measurements are synchronised with the simulation study using BEM modelling combined with the optimisation approach.

7 Identification of corrosion location and area in operation

7.1 Method of identification of corrosion location and area

New corrosions areas can be created during ship operation, specially on the highly turbulent areas, such as the areas close to the propellers. If the location

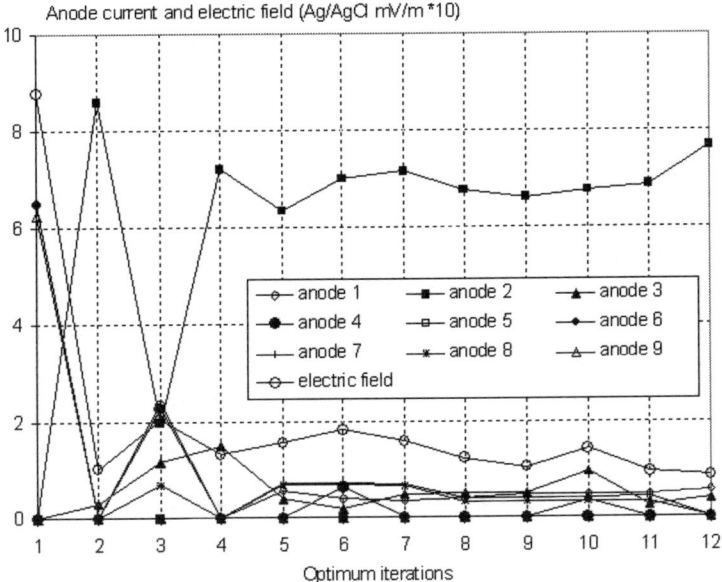

Figure 29: Optimisation iterations in measurement process, scaled seawater in static.

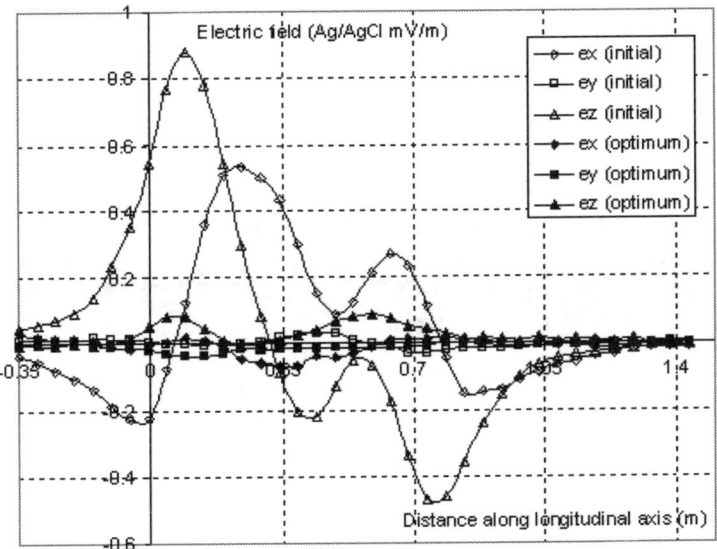

Figure 30: Electric fields before and after optimisation, scaled seawater in static.

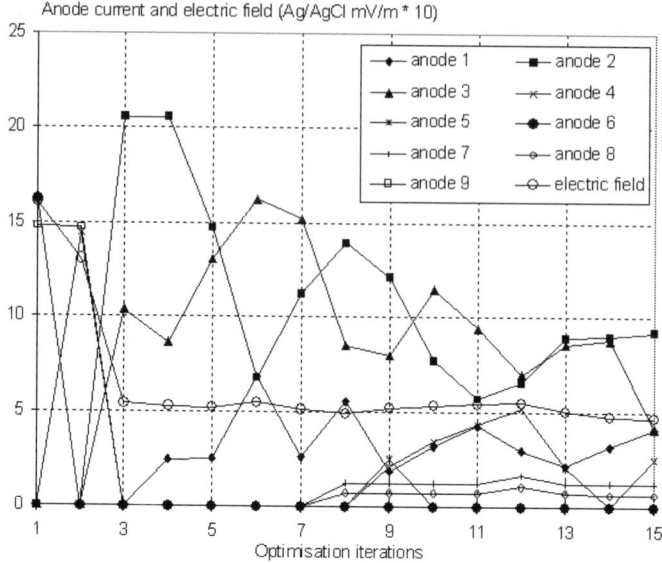

Figure 31: Optimisation iterations in measurements, simulated dynamic corrosion condition.

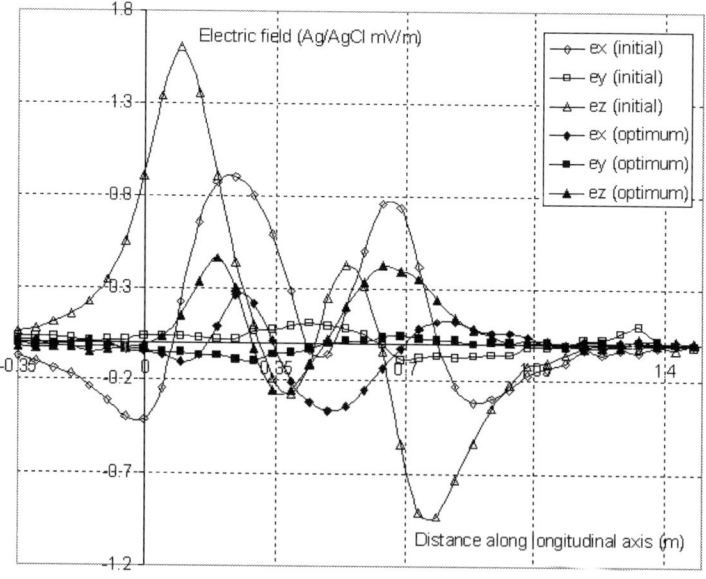

Figure 32: Electric fields before and after optimisation, simulated dynamic corrosion condition.

and strength of the new corrosions can be identified during ship sailing, it might be possible to reconfigure the ICCP currents to reduce the electric signature.

The identification of the location and strength of the new corrosions is an inverse problem. We propose a solution to the inverse problem by defining a set of pitches on the ship hull to present the corrosion and treating each corrosion pitch as a unit corrosion area. The potential produced by the unit corrosion area can be simulated via BEM modelling. If the potentials corresponding to new corrosions can be measured, the inverse problem may be stated as,

$$
\begin{bmatrix} \hat{u}_1^m \\ \vdots \\ \hat{u}_n^m \\ \vdots \\ \hat{u}_N^m \end{bmatrix} = \begin{bmatrix} u_{1,1} & \cdots & u_{1,m} & \cdots & u_{1,M} \\ \vdots & & & & \\ u_{n,1} & \cdots & u_{n,m} & \cdots & u_{n,M} \\ \vdots & & & & \\ u_{N,1} & \cdots & u_{N,m} & \cdots & u_{N,M} \end{bmatrix} \begin{bmatrix} u_1^c \\ \vdots \\ u_m^c \\ \vdots \\ u_M^c \end{bmatrix} + \begin{bmatrix} \varepsilon_1 \\ \vdots \\ \varepsilon_n \\ \vdots \\ \varepsilon_N \end{bmatrix} \tag{20}
$$

or,

$$
\begin{bmatrix} U^m \end{bmatrix} = \begin{bmatrix} A \end{bmatrix} \begin{bmatrix} U^c \end{bmatrix} + \begin{bmatrix} \varepsilon \end{bmatrix} \tag{21}
$$

where N is the total number of the selected measurement points of the corrosion protection potentials on the ship hull, M is the total number of the unit corrosion areas on the ship hull, u_m^c is the corrosion potential produced by m^{th} corrosion area, A is a transfer matrix and its component $u_{n,m}$ is defined as the unit potential created by m^{th} corrosion area on the measurement points, $\hat{u}_n^m$ is the measured potential on the measurement points, ε is the measurement error.

7.2 Objective functions and constraints of the inversion

The objective function and constraints of the inverse problem are defined as:

$$
\text{Minimise} \quad f_u = \left(\sum_{m=1}^{M} u_{n,m} \, \hat{u}_m^c - u_n^m \right)^2 \qquad n = 1, 2, \ldots, N \tag{22}
$$

$$
\text{Subject to:} \quad \begin{cases} u_{\min} \le u_n^m \le u_{\max} \\ \hat{i}_k \ge 0 \end{cases} \qquad k = 1, 2, \ldots, K \tag{23}
$$

where $u_n^m = (u_1^m, u_2^m, ..., u_N^m)$ are the corrosion protection potentials at the measurement points produced by new corrosion condition, $i_k = (i_1, i_2, \cdots, i_K)$ are the anode currents and K is the number of anodes, u_m^c is the corrosion potential produced by m^{th} corrosion area, $\hat{u}_m^c$ and $\hat{i}_K$ are the estimators of u_m^c and i_k, respectively.

The least square approach to (22) is to solve:

$$\text{Find} \quad U^c \quad \text{such as} \quad \left\| A\, U^c - U^m \right\| \quad \text{to be minimum} \qquad (24)$$

In general, the equation numbers obtained from measurements usually are much more than unknowns. If two or more measurements to fit data equally well or equally bad (the measurements are inaccurate with higher noise and uncertainty), matrix [A] is unable to distinguish between them. The matrix [A] folds up its tent and becomes the singular. The problem is ill-posed and there is no guarantee of the uniqueness of the solution. To avoid the singular problem in the normal equation, Singular Value Decomposition (SVD) method is often used in the inverse problem.

Furthermore, it is common to use a statistical formulation and regularisation methods in order to limit the space-parameters zone. The solution is defined as the set of parameters possibly closer to reference values presented as the initial or the previous estimation. The most commonly used methods are Tikhonov's, Levenberg's, and Levenberg-Marquardt's. All the methods introduce a regularisation term to represent the least square difference between the calculated M and the initial guessed one or the previous calculated one. The formulation using Tikhonov's [32] is,

$$T^\alpha \left[U^c \right] = \left\| A U^c - U^m \right\|^2 + \alpha \left\| U^c \right\|^2 \qquad (25)$$

where $\alpha > 0$ is the regularisation parameter that should be chosen.

Both SVD and stochastic methods were tested. The inaccurate inverse solution using SVD was found when the differences between corrosion protection potentials produced by new corrosions and original corrosion protection potentials are very small. In the use of stochastic method SA, we have introduced a factor to accelerate the convergence speed and the objective functions are modified as,

$$\text{Minimise} \quad f_u = \left(\sum_{m=1}^{M} u_{n,m}\, \hat{u}_m - u_n^c \right)^\xi \quad n = 1, 2, ..., N \qquad (26)$$

ξ is a parameter need to be adjusted to achieve a best convergence for the inversion.

7.3 Inversion results

An investigation was conducted using the BEM model, same as fig. 10, to verify the proposed solution to the inverse problem.

There a total of 20 potential measurement points are arranged along the keel line of the ship hull. Corrosion areas c5 and c6 on the ship hull, as illustrated in fig. 10, were selected and assumed to have been corroded completely.

Only one anode was selected and is fixed at anode position a2 as illustrated in fig. 10. Potentials produced by each corrosion area were simulated using BEM modelling. The potentials produced by each corrosion area (c1 to c10) measured along the keel line are illustrated in fig. 33. These potentials were treated as the unit potentials in the inversion process.

Each individual corrosion area creates a potential anomaly. The inversion is to match the potential anomaly by finding the optimum combinations of the numbers and amplitudes of the unit potentials based on objective functions (26) and constraints (23).

The iteration processes are illustrated in fig. 34 where the SA starts in an average value of unit corrosions. Parameter ξ was tested preliminarily and fixed to equal to 2.52.

The final results of the inversion are listed in table 8. The inversion results show that the corrosion locations and strengths are identifiable. The inversion error is generally small as shown in fig. 35.

The computed potentials by BEM modelling and inverted potentials for the new corrosion areas c5 and c6 are illustrated in fig. 36. The difficulty of the inversion is in defining the unit potential to match perfectly to the changes of the corrosion areas since the anomaly of the potentials caused by the new corrosion is generally small in amplitude.

Table 8: Inverted corrosion areas and strengths.

Unit corrosion area	New corrosion area (%)	Inverted corrosion area (%)	Inversion error (%)
1	0.0	0.0	
2	0.0	0.0	
3	0.0	0.0	
4	0.0	0.61	
5	100.0	95.65	-4.35
6	100.0	102.21	+2.21
7	0.0	2.22	
8	0.0	0.0	
9	0.0	0.0	
10	0.0	0.0	

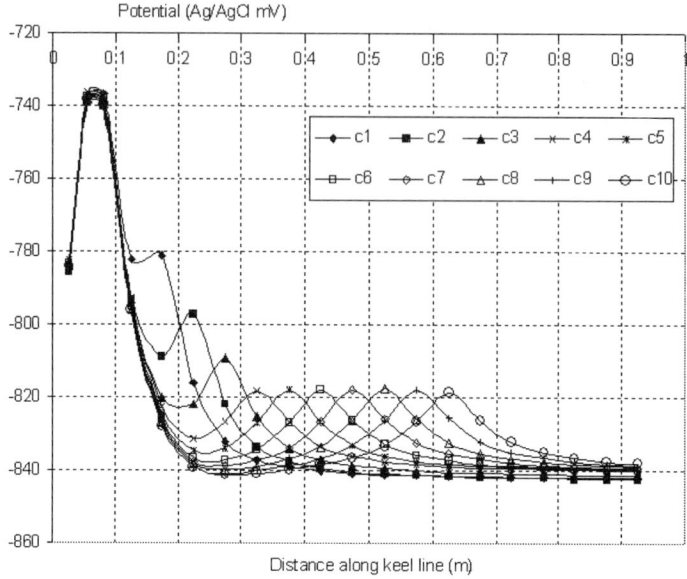

Figure 33: Potentials generated by unit corrosion area.

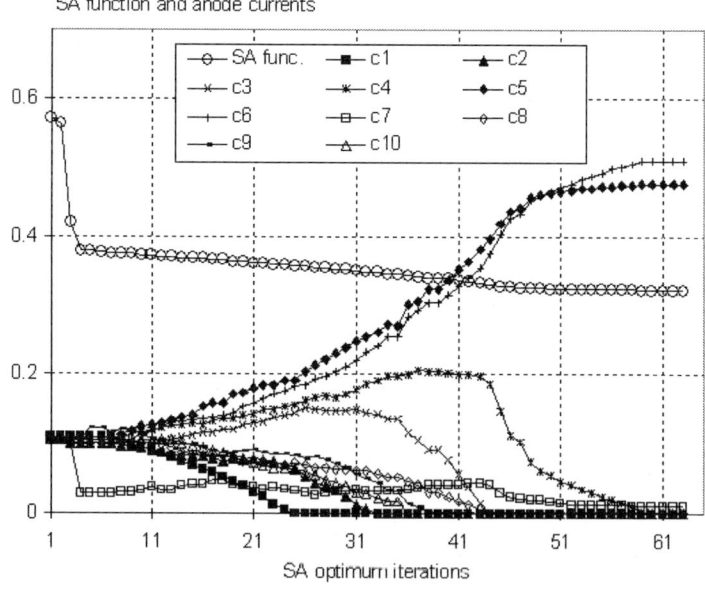

Figure 34: Optimum SA function iterations (SA objective function are given using $\log_{10}(SA\ Objective\ function)/10$).

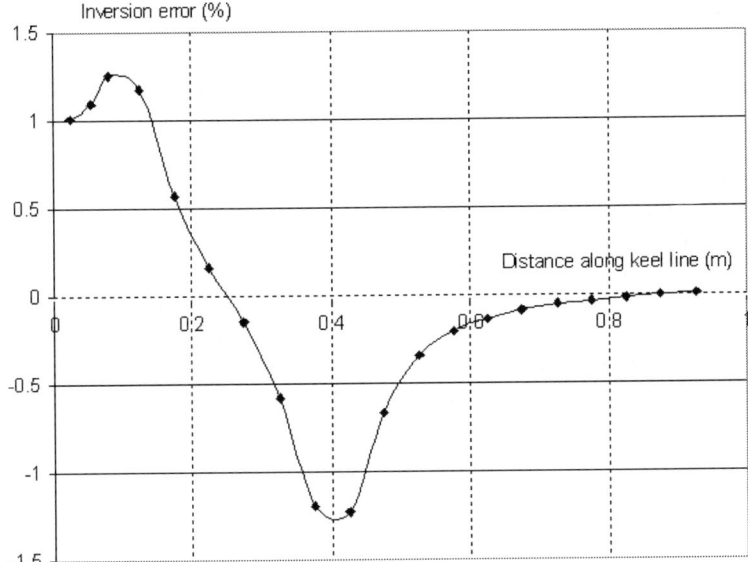

Figure 35: Potentials inversion error.

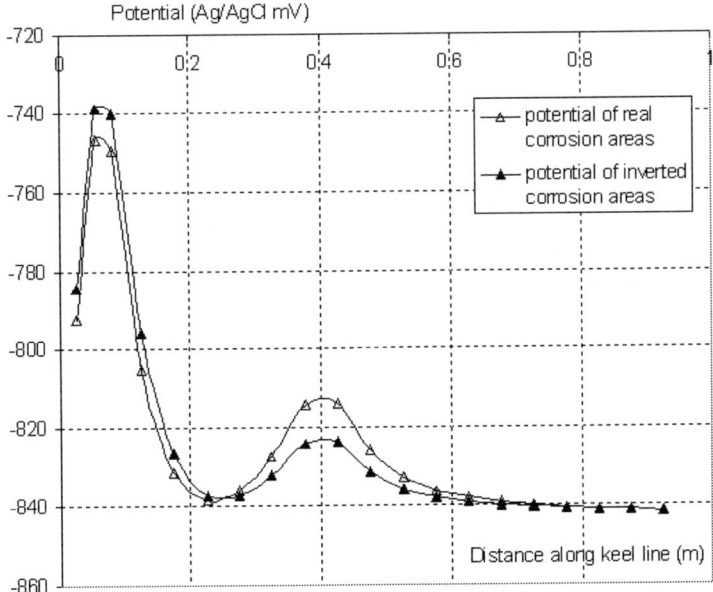

Figure 36: Potentials by BEM modelling and by inversion.

8   Conclusions

The joint optimisation approach of corrosion protection and electric signature of a naval platform is presented. The results show that the electric signature could be reduced with a sufficient corrosion protection and this could be achieved by a combination approach using boundary element modelling, optimisation and experimental techniques.

The corrosion modelling to a ship is a difficult task due to the complex structures on the wetted surface of the ship hull, corrosion conditions, interactions of corrosion materials, influences of hydrodynamics and environments. Simplifications have to be made in BEM model and linearity material polarisation curves might be utilised. Though the optimisation results are obtained with these constraints, the optimisation might be able to provide a useful reference to the optimisation design and ranging processes of a naval platform.

Identification of new corrosion areas and strengths during ship sailing in seawater would benefit operation since the electric signature could be monitored and reduced in real time. This might be possible through an inversion process. The present inversion to the corrosion changes is only a simplification to the real operational environments and more studies will be needed in the future.

Acknowledgement

The author would like to thank Dr J. T. Goh for his support and suggestions during this work.

References

[1]   Hubbard, J.C., Brooks, S.H. & Torrance, B.C., Practical Measures for the reduction and management of the electro-magnetic signatures of in-service surface ships and submarines, UDT 96, Jul. 1996.
[2]   Holt, Richard, Detection and measurement of electric fields in the marine environment, UDT 96, Jul. 1996.
[3]   Duncan, S.M. & Webb, G., Multi influence transportation signature range, *Proceeding IMDEX ASIA* 1999.
[4]   Pei, Y.H. & Abey, R.A., Design optimisation of ship ICCP system, BEM 21, Oxford, UK, Aug., 1999.
[5]   Pei, Y.H. & Adey, R.A., Reduction of Ship's Electric and Corrosion Related Magnetic Signatures Using Optimisation Techniques Combined with Numerical and Analytical Modelling, 3rd Marine elecromagnetics, Stockholm, Sweden, July 2001.
[6]   Satana Diaz, E., Baynham, J., Adey, R. & Pei, Y.H., Optimisation of ICCP systems to minimise electric Signatures, 3rd Marine elecromagnetics, Stockholm, Sweden, July, 2001.

[7] Johnson J. Michael & Yahya, Rahmat-Samii. Genetic algorithm in engineering electromagnetics, *IEEE Antennas and Propagation Magazine*, **39(4)**, 1997.

[8] Preis, K., Magele, C. & Biro, O., FEM and evolution strategy in the optimum design of electromagnetic devices, *IEEE Trans. on MAG*, **26(5)**, 1990.

[9] Alotto, Pier Giorgio, *et al*, Stochastic algorithm in electromagnetic optimisation, *IEEE Trans. on MAG*, **34(5)**, 1998.

[10] Simkin, J. & Trowbridge, C.W., Optimizing electromagnetic devices combining direct search method with simulated annealing, *IEEE Trans on MAG*, **28(2)**, 1992.

[11] Brebbia, C.A. & Dominguez, J., Boundary elements: an introductory course, CML Publications, Southampton, 1980.

[12] DeGiorgi, *et al*, Boundary element evaluation of ICCP system under simulated service conditions, *Boundary Element Technology VII*, 1992.

[13] Zamani, *et al*, A survey of computational efforts in the field of corrosion engineering, *Int. J. Numerical Methods in Engrg.*, **23**, 1986.

[14] Finoly, G., *et al*, Application and validation of boundary element method to cathodic protection designs on vessels, *Boundary Element Technology VIV*, 1992.

[15] Strommen, R.S., Computer modelling of offshore cathodic protection systems: method and experience, *Computer Modelling in Corrosion*, ASTM STP 1154, Philadelphia, 1992.

[16] Abey, R.A. & Pei, Y.H., Computer simulation as an aid to corrosion control and reduction, NACE, *Conference 99, San Antonio, Texas, USA, 1999.*

[17] Computational Mechanics, BEASY-CP User Guide, *Computational Mechanics BEASY, Aug. 2000.*

[18] Pei, Y.H., DSO report, UWP/01490/2000, Aug. 2000.

[19] Weaver, J.T., The quasi-static field of an electric dipole embedded in a two-layer conducting half-space, *Canada Journal of Physics*, **45**, 1967.

[20] Kaesche, H., Metallic Corrosion, NACE, Houston, Texas, 1985.

[21] Francis, L. Laque, Marine Corrosion, John Wiley & Sons, 1977.

[22] Plapp, J.E., Engineering fluid mechanics, Prentice-Hall, New Jersey, 1986.

[23] Harvey, P. Hack, Atlas of polarisation diagrams for naval materials in seawater, Carderock Division, Naval Surface Warfare Centre, Apr. 1995.

[24] Fletcher, R., Practical methods of optimisation, Wiley, New York, 1987.

[25] Davis, M. Himmelblau, Applied nonlinear programming, McGraw-Hill, 1972.

[26] Aarts, E. & Korst, J., Simulated annealing and Boltzmann machines, John Wiley & Sons, 1989.

[27] Pei, Y.H., Investigating corrosion protection design of waterjet propulsion using boundary element technology, *Boundary Element Technology XIII, 1999.*

[28] Ditchfield, R.W., McGrath, J.N. & Tighe-Forde, D.J., Theoretical validation of the physical scale modelling of the electric potential characteristics of marine impressed current cathodic protection, *Journal of Applied Electrochemistry*, **25**, 1995.

[29] Hack, H.P. & Scully, J.R., Galvanic Corrosion protection using long- and short-term polarisation curves, Corrosion, **43**, NACE, USA, 1996.

[30] Tighe-Ford, D.J., Bottern, R.A. & Hughes, R.D., Study of design criteria for ship impressed current cathodic protection by stylised modelling, *Corrosion Prevention & Control*, Feb. 1990, UK, 1990.

[31] Hack, H.P., Scale Modelling for corrosion studies, *materials performance*, 1989, NACE, USA, 1989.

[32] Tikhonov, A.N., Goncharsky, A.V., Stepanov, V.V. & Yagola, A.G., *Numerical methods for the solution of ill-posed problem*, Dordrecht / Boston / London, Kluwer Academic publishers, 1995.

Optimisation of cathodic protection systems

E.S. Diaz & R.A. Adey
C.M. BEASY Ltd., UK.

Abstract

Cathodic protection is used to prevent corrosion of the hull of a marine vessel to avoid thinning of the hull plates, damage to welds and initiation of pitting, and thus reducing the maintenance on the ship and preserving the smoothness of the hull. Boundary Elements Methods (BEM) has been widely used in the cathodic protection field to simulate the performance of the CP system and to predict the associated corrosion related electric and magnetic fields surrounding the vessel [1]. In this work optimisation methods are developed which combine with the BEM to predict the minimum currents of the anodes to protect the vessel within certain range of protection potential. The global optimisation approach enables corrosion engineers to define the objectives of the simulation rather than predicting the response to a specific set of conditions.

1 Introduction

The protection of a metal is achieved when the potential of the structure relative to the corrosion cell boundary is polarised to be more negative than the open circuit potential of the anode [2].

In practice, there are two forms of criteria that can be employed to determine the degree of the cathodic protection by measurement of the potential of the structure.

- With certain metals, there is a maximum anode open circuit potential in a particular electrolyte and the potential of the structure relative to a standard half cell placed at the corrosion cell boundary can be used to determine the adequacy of protection.
- As an alternative, it is sometimes possible to specify a maximum potential difference that will exist between the anode open-circuit and the corrosion cell potentials. Protection is achieved by causing a potential change of the

metal relative to the corrosion cell boundary, greater than this potential difference.

The first criterion has been widely used by several authors and it is employed in this work as a criterion for determining the degree of cathodic protection.

Cathodic protection can stop the corrosion of the hull so that the hull plates are not thinned, the welds not consumed and pitting is avoided. This will reduce the maintenance on the ship. Alternatively, cathodic protection can be used to maintain the hull's smoothness. In a normal sample of scale from an unprotected ship there is a sandwich of layers of paint and rust, which can lead to a roughening of the hull.

It is possible during cathodic protection to supply excess direct current to polarise a structure below the recommended protection potential. This state of affairs is termed '*overprotection*'. There are two main consequences of over-protection, namely, waste of current and more seriously the violation of the structural integrity of the metal. The waste of current is due to the polarisation of the metal below its equilibrium potential with the excess current being used to evolve hydrogen. The gas produced could cause the detachment of organic coatings and the removal of calcareous deposits in offshore structures. Hydrogen production has also adverse effects on both the corrosion fatigue life and hydrogen embrittlement properties of structures particularly those made of high strength materials. On bare surfaces immersed in seawater, these could have a beneficial effect since the hydroxyl species may passivate and/or enhance the formation of calcareous deposits which in turn will reduce the current demand. However, for organically coated surfaces the strong alkali condition at the metal surface may result in loss of adhesion for the paint. This phenomenon is known as cathodic disbonding.

Preventing coating damage by excessive cathodic protection is best accomplished by avoiding over-protection. The polarisation potential should be kept below the hydrogen over-voltage potential, which is the point at which free hydrogen starts to evolve. For example, free hydrogen evolution on steel pipe can be looked for when the polarisation potential approaches a value in the order of $-1.2V$ as measured to a copper-copper sulphate reference electrode.

In summary, the potential of the metal surfaces, particularly steel, requires to be constrained to a target range: superior value to protect the surface from under protection, and inferior value to avoid overprotection. These constraints have been tried by the technique of 'trial and error'. This technique could give some reasonable solutions but it will certainly not give the best one. Furthermore, reaching a point where these constraints are satisfied can take considerable time, much more than normally available. In addition, when the current density that satisfies the constraints is reached, the current is so high that the average potential of the surface is far away from the objective initially set up. Simple design formulas based on the area to be protected and the current required per unit area provide some initial estimates, but they can lead to over protection or under protection in all but the simplest situations. Therefore there is a need for a more accurate method of designing CP systems, which can consider all the factors such as the geometry, environmental conditions and the materials [3–5].

An optimisation process could help to find the correct solution within the constraints [6]. Automatic design optimisation provides, in a computer simulation, the key data on the required CP system to protect the structure. For example, the current at the anode which protects the structure, size of the anode, number of anodes and so on. In this work, only the current at the anode is analysed by using two objective functions.

The declared objectives are:

- Minimise the anode current. The solution of this case will give the minimum necessary current coming out from the anodes which satisfies the constraints. This, however, could create a non uniform potential over the hull surface which would lead to a possible roughening of the hull.
- Achieve a uniform potential over the surface of the structure. The solution of this case will give the optimum current coming out from the anodes which satisfies the constraints and distributes the potential over the surfaces as uniformly as possible. This it to say, that the potential difference among the elements on the surface is the smallest possible (uniform).

These optimisation objectives are mainly applicable to impressed anodes, other cases more applicable to sacrificial anode systems are reported elsewhere [7–9].

1.1 State of the art

In 1987, N.G. Zamani and J.M. Chuang studied the optimal control of current in an impressed cathodic protection system [10].

In their work, since the polarisation curve was treated as linear, the objective function was included in the resolution of the Laplace equation. They tried to determine the current output as an unknown control to be determined by posing the following question: find the current output $i = -u$ on Γ_A, such that the potential distribution on the cathode Γ_C minimises the objective function

$$J[\phi, u] = \int_{\Gamma_C} (\phi - \phi_t)^2 \, d\Gamma \tag{1}$$

The eqn (1) was slightly modified to introduce the anode current in another term.

$$J[\phi, u] = \frac{1}{2} \int_{\Gamma_C} (\phi - \phi_t)^2 \, d\Gamma + \frac{\alpha}{2} \int_{\Gamma_A} u^2 \, d\Gamma + \tag{2}$$

Where:
ϕ is the potential.
u is the current density.

ϕ_t is the target potential above which the surface is corroded.
Γ_C cathode surface.
Γ_A anode surface.
α is a design weighting factor.

Several 2D models were analysed. The solution obtained with this method would give the optimum current density which set the surfaces as close as possible to the target value.

Since the solution obtained by N.G. Zamani and J.M. Chuang did not guarantee the complete protection in which the potential values on the whole surface of the metal are kept below the critical value ϕ_t. K. Kishimoto *et al* [11, 12] (1990) did further analysis in which the objective function minimised was:

$$P = \int_{\Gamma_e} \phi u d\Gamma \tag{3}$$

Γ_e anode surface.
Subject to the following constraint on the surface of the cathode:

$$-\phi \le \phi_p \text{ on } \Gamma_C \tag{4}$$

Therefore, the final solution obtained was guaranteed to be within the required potential range and also be the minimum value of the current at the anodes.

The complete objective function was defined as:

$$P^* = P + k \left\{ \|r\|^2 + \|h\|^2 \right\} \tag{5}$$

Where:
K is the penalty coefficient.
r is the left hand side of (6):

$$k[H] \left\{ \begin{matrix} \phi_0 \\ \dots \\ \phi_n \\ \dots \\ -f(q_m) \\ \dots \\ \phi_e \end{matrix} \right\} - [G] \left\{ \begin{matrix} q_d \\ \dots \\ q_0 \\ \dots \\ q_m \\ \dots \\ q_e \end{matrix} \right\} = \{0\} \tag{6}$$

Based on the boundary conditions applied:

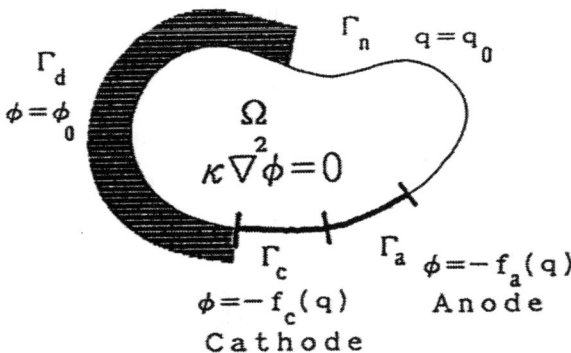

Figure 1: Boundary conditions of galvanic corrosion problem.

m is the metal.

f(qm) is the value of the potential taken from the polarisation curve for the metal.

$$h_i = (-\phi_i + \phi_p) \, u \, (-\phi_i - \phi_p) \quad (I = 1, 2, \ldots, K) \tag{7}$$

Where:

u($\cdot$) is the unit step function.

ϕ_i is the potential at the nodal point i

K is the total number of nodes on the surface of the hull.

ϕ_p is the critical value of potential for protection needed to satisfy.

q is the current density.

In their work, the polarisation curve was arranged following the Tafel equations [13]:

$$-\phi_A = 1 \pm \log |u| \tag{8}$$

$$-\phi_B = 1 \pm \log |u| \tag{9}$$

Where:

ϕ_A is the anodic potential.

ϕ_B is the cathodic potential.

Aoki and Amaya [14] (1992) took into account the polarisation of the metal to be protected.

Hou and Sun [15] (1994) adjusted the current density and the location of the anodes and the objective function employed was the objective function previously used by Zamani and Chuang in eqn (2). The polarisation curve applied was the empirical function given by the Butler–Volmer equation, two approaches to this equation were assumed:

- The polarisation on the cathode is small, therefore the Butler–Volmer equation can be assumed as being linear:

$$F(\phi) = \tau_0 + \tau_1 \phi \tag{10}$$

Where τ_0 and τ_1 are constants.
- Or a non-linear Butler–Volmer equation:

$$F(\phi) = -k\phi - l\phi^3 \tag{11}$$

Where k, l are positive constants.

Aoki *et al* [16] (1996) also determined the optimal currents to be impressed from each electrode by defining a cost function:

$$P^*(Q) = P(Q) + \upsilon \sum_{k=1}^{K} h_k^2(Q) \tag{12}$$

Where $P(Q)$ is the power necessary for cathodic protection given by:

$$P(Q) = \sum_{e=1}^{N} \int_{\Gamma_e} \left\{ \phi_e(Q) + f_e(i_e(Q)) \right\} i_e(Q) d\Gamma \tag{13}$$

Q is a vector which components are i_e (e = 1, 2,..., N), currents of the electrodes.

ϕ_e potential at the electrodes.

The polarisation characteristics of the electrode-e is given by a known function $-\phi = f_e(i)$.

N is the number of electrodes (anodes).

υ is a penalty number (large position)

$$h_k = (-\phi_k + \phi_p) \, u \, (-\phi_k + \phi_p) \; (k = 1,2, \dots, K) \tag{14}$$

Where:

$u(\cdot)$ is the unit step function.

ϕ_k is the potential at the nodal point k.

K is the total number of nodes on the surface of the hull.

ϕ_p is the critical value of potential for protection needed to satisfy.

If the protection condition is not satissfied, $P^*(Q)$ is not satisfied, it takes a large value.

Hand and Adey [17] presented an approach to the global optimisation of a ship's ICCP system coupling the boundary element method with the global optimisation method 'Simulated Annealing'. The anode was moved from element to element on the search area. The objective function employed has also applied a penalty term:

$$P = \left(\sum_{k-1}^{n} \frac{|u_t|}{|u_t| - |u_t - u_k|} \right) - \frac{\varepsilon}{P(i,u,s)} + r \sum \frac{1}{G(i,s,u,P)} \qquad (15)$$

Where:

$$\text{with } u_{min} \le u_k \le u_{max} \quad k = 1,\ldots,n \qquad (16)$$

$P(i,u,g)$ is the electric power consumption on the anodes.

ε is a weight factor to incorporate the power consumption in the objective function.

i is the anode electric currents.

u is the potential at the measurement points.

s is the location of the anodes.

u_{min} and u_{max} is the minimum and maximum protection potential level on the wetted surface of the ship.

$u_k = (u_1, u_2,\ldots, u_n)$ is the potential at the measurement points, which are required to satisfy the constraint u_{min}

u_t is the threshold potential and is defined on the reference cell.

r is a positive weighting coefficient which ensures that during the optimisation the constrains $G(i,s,u,P)$ are not violated.

The constraints are:

$$G(i,s,u,P) = \left(\sum_{k=1}^{n} u_k - \sum_{j=1}^{N} u_j \right) \ge 0 \qquad (17)$$

Where the second term is a defined threshold potential and N is a threshold number of measured potentials which meet $u_{min} \le u_j \le u_{max}$.

In this work, two methods of controlling the potential over the metal surfaces are introduced, the first one, minimising the current at the anodes constraining the potential on the surfaces to the protection range, the second one, achieving a uniform potential at the surfaces around a target value satisfying the constraints. The optimisation methods employed in the previous work is the unconstrained multivariable search in which a penalty factor was included to control the potential within the range. Constrained multivariable methods are applied in this work since the potential is required to be within a certain range. In this way, the results are always assured to be within the protection range not open to possible designs which could not satisfy this range. The optimisation methods used are the Modified method of Feasible Directions (MFD), Sequential Linear Programming (SLP) and Sequential Quadratic Programming (SQP).

In this work, the performance of the two methods of controlling the potential over the surfaces are studied (*Minimise the anode current density, Achieve a uniform potential over the surface of the structure*). In addition, three methods of optimisation are used and compared to gain conclusions of which one yield to the best results. All of these experiments are conducted using a model of a frigate model and submarine as described below.

Figure 2: Frigate model. The dimensions of the frigate are:
  – Waterline length: 34.0m.
  – Draft: 2.3m.
  – Waterline beam: 6.4m.

1.2 Frigate model, description

The model has the characteristics shown in fig. 2. The frigate was considered fully coated apart from some specific damaged areas. The propeller was set made of nickel-aluminium-bronze.

To speed up the solution and since the model is symmetric only half of it was modelled.

The electrolyte considered was seawater with a resistivity of 20ohm·cms [2], which implies a conductive of about 5S/m.

The frigate was modelled with 1338 elements in most of the experiments analysed. A more refined mesh was created at the stern of the vessel since it is the most critical area of the frigate due to the propeller and location of anodes.

1.3 Submarine model

A model of a submarine was also considered to validate the conclusions obtained with the frigate model. The model has the characteristics shown in fig. 3. The submarine was considered to be fully coated apart from some specific damaged areas. The propeller was set made of nickel-aluminium-bronze. Damaged areas were placed along the hull surface in areas more likely to receive damage during the life of the ship.

The electrolyte considered was seawater with a resistivity of 20ohm·cms [2], which implies a conductive of about 5S/m.

The model has 1087 elements including the surrounding box, which simulates the electrolyte.

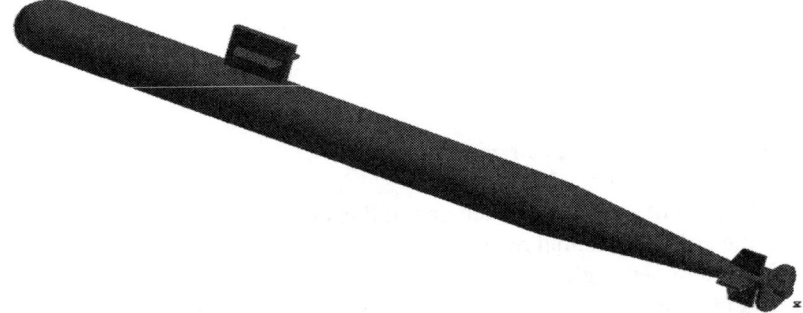

Figure 3: Submarine model. The dimensions of the submarine are:
- Waterline length: 100.0m.
- Beam: 8.0m.

1.4 Minimising the current at the anodes

In this case, the objective is to minimise the total anode current while at the same time ensuring that the protection potential on the surface of the vessel is within the required range to avoid under or over protection. This process automates the task of the corrosion engineer who normally has to use his experience and a process of trial and error to achieve the same result.

- The *summation of the currents* is used as objective function to minimise the current at the anodes.

$$Obj = \sum_{i=1}^{n} I_i \tag{18}$$

The objective will simply minimise the total current irrespective of the individual anode values. This could result in the current of a particular anode being reduced to zero indicating that it is not required.

- The *least squares of the currents* (19) will tend to results where the anode currents are fairly uniform clustered around a minimum. This is shown clearly in the first example of this work where eqns (18) and (19) are compared.

$$Obj = \sum_{i=1}^{n} I_i^2 \tag{19}$$

Subject to the following constraints on the surface of the cathode:

$$g_i = \frac{\phi_i - \phi_{Max,i}}{|\phi_{min,i}|} \leq 0 \text{ on } \Gamma_c \quad i = 1,\ldots,m \tag{20}$$

$$g_j = \frac{\phi_{min,i} - \phi_i}{|\phi_{min,i}|} \le 0 \text{ on } \Gamma_C \quad j = 1,\dots,m \tag{21}$$

Where:

Γ_C is the surface of the cathode.

n is the number of anodes.

I is the current supplied by each one of the anodes.

ϕ_i is the current potential over the element i.

m is the number of elements in the cathode.

ϕ_{min} is the minimum potential required to avoid overprotection.

ϕ_{max} is the maximum potential required to avoid underprotection.

Note. It is normally recommended for the optimisation algorithms [18] that to obtain a well-conditioned problem the constraints will have roughly the same order of magnitude as the gradients (within a factor of 100). Therefore the denominator of eqns (20) and (21) were applied to scale the constraints into non-dimensional units.

Since constraints are considered, the final solution obtained is guaranteed to be within the required potential range. Several experiments were analysed using three methods of optimisation.

1.5 Numerical examples, minimise the current at the anodes

Without a good deal of experience, the corrosion engineer does not have a concrete idea of the size, the number, or even the current to be supplied by the anodes. The optimisation of the current at the anode provides the correct current to protect the areas of the structure and consequently gives the information to dimension the real anodes to be set on the structure.

In sea water, the most common reference cell is the silver chloride cell (Ag/AgCl/seawater) and the established potential criterion for the protection of steel is −800mV. If the reference cell is at this value of potential the nearby surfaces will also be protected. The purpose of these experiments is to dimension the anodes by optimising their currents. The potential of the damaged surface of the structure was required to be within a certain potential range −850mV and −1100mV in order to ensure that the structure is protected. This will ensure that the design is sufficient to protect the structure.

Several damaged areas were placed along the vessel hull (1.2) as it is shown in fig. 4.

Since the damage is moderately significant, the three areas at the bow, the keel and the stern are assumed to have a similar behaviour to the bare steel. Their areas are shown in table 1.

The propeller was set made of Nickel-Aluminium-Bronze with an area of 1.84m^2.

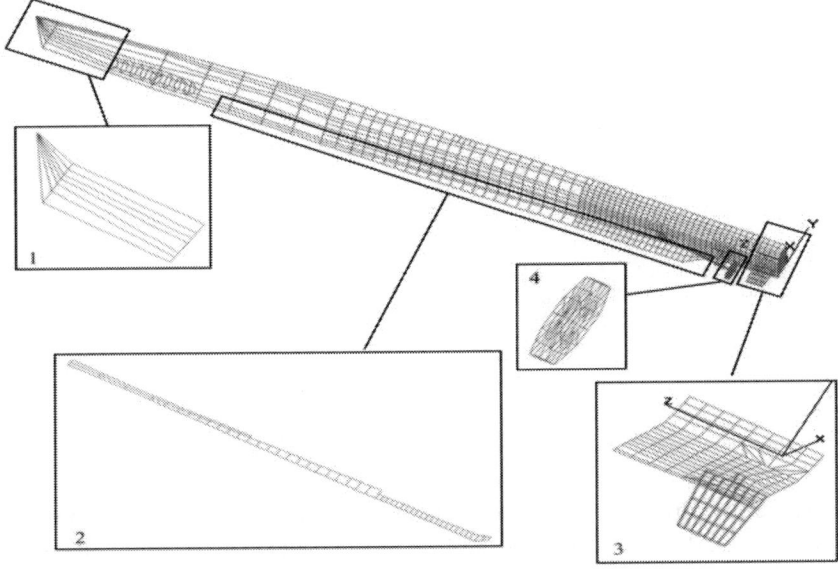

1.- Damaged surface at the bow of the ship, bare steel. [cpopt metal surface bow]
2.- Damaged surface at the keel of the ship, bare steel. [cpopt metal surface keel]
3.- Damaged surface around the rudder of the ship, bare steel. [cpopt metal surface stern]
4.- Propeller, nickel aluminium bronze. [cpopt metal surface propeller]

Figure 4: Propeller and damaged areas on the ship hull.

The frigate model was considered to have six anodes. Since the model is symmetric, only half of the model was studied. Three anodes were placed in a symmetric section of the hull of the vessel in order to protect the damaged areas; one was placed near the bow, the second one in the centre of the hull and the third one at the stern of the ship (fig. 5).

For simplicity, the anode area is smaller than a real size anode. It is assumed that its performance is similar to a real size anode. The areas are shown in table 2:

The initial currents and current densities applied to each one of the anodes are shown in table 3:

The optimisation was launched using the Sequential Linear Programming method and the objective to minimise the total current (eqn 18).

Table 1: Area of the damaged areas.

Position	Name	Area(m²)
Area at the bow of the vessel	cpopt metal surface bow	4.34
Area at the keel zone	cpopt metal surface keel	7.96
Area around the stern	cpopt metal surface stern	8.01

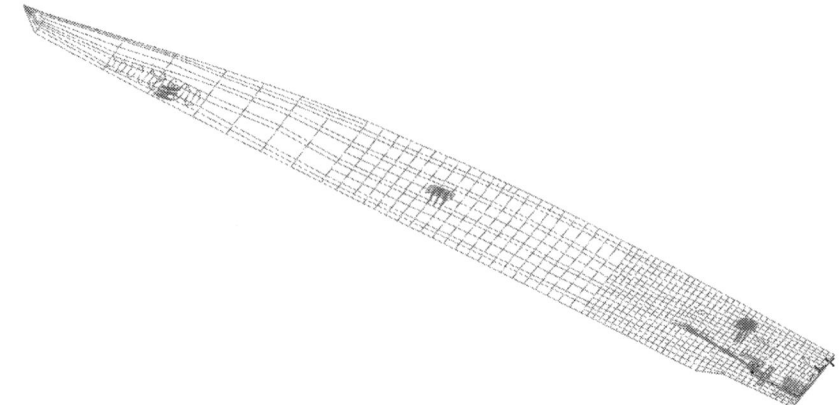

Figure 5: Anodes distribution on the ship hull.

The initial distribution of potentials over the surface had an under protected profile (fig. 6) since the potentials in the damaged areas were higher than −850mV. After the optimisation, the current at the anodes was increased to reach the potential range required:

The results of the optimisation highlighted that in order to minimise the current at the anodes, the anode at the keel should be deactivated, and the anode at the stern should supply most of the current. As pointed out earlier this is possible using eqn (18).

Equation (19) yielded the results shown in table 5 where *the current at the anodes have been clustered around a minimum value*, due to the least squares approximation. Consequently, the minimum total current supplied by the anodes was not found due to the least squares definition of the objective function. As a result, this equation was considered inadequate to the purpose of reducing the amount of current supplied by the anodes.

Table 2: Area of the anodes.

Position	Name	Area(m^2)
Anode at the bow of the vessel	cpopt anode bow	0.10
Anode at the keel zone	cpopt anode keel	0.10
Anode at the stern	cpopt anode stern	0.05

Table 3: Initial currents and current densities applied to the anodes.

Name	Current Density (mA/m^2)	Current (mA)
cpopt anode bow	−3000	−315.3
cpopt anode keel	−3000	−300.1
cpopt anode stern	−3000	−159.4

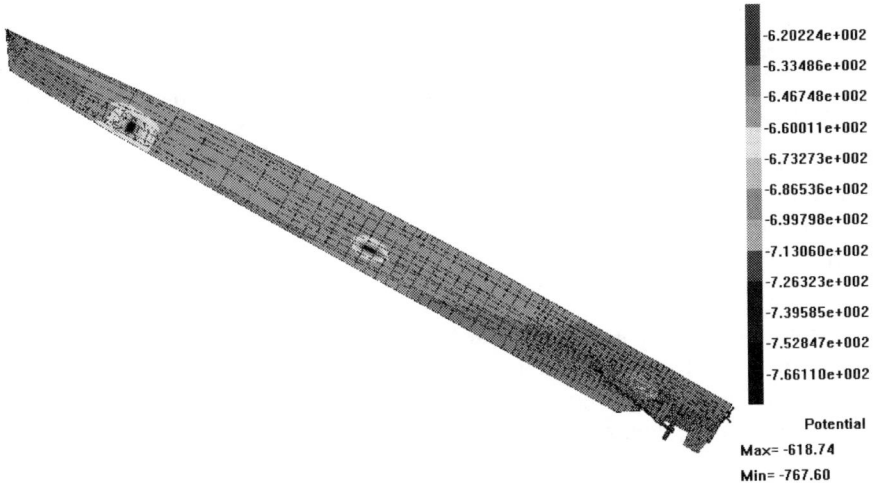

-6.20224e+002
-6.33486e+002
-6.46748e+002
-6.60011e+002
-6.73273e+002
-6.86536e+002
-6.99798e+002
-7.13060e+002
-7.26323e+002
-7.39585e+002
-7.52847e+002
-7.66110e+002

Potential
Max= -618.74
Min= -767.60

Figure 6: Potential profile of the hull of the ship for the initial currents supplied.

The computations (table 5) were carried out using the eqn (18).

The number of elements in the range of the potential required (−850mV, −1100mV) is zero at the beginning of the process since none of the damaged surfaces have their potential in the required range (table 6).

The number of elements in the range of the potentials is increased until all the areas are protected, during the optimisation process. The objective function value is the summation of the currents supplied by the anodes: 7486.3mA (fig. 7).

Table 7 and fig. 8 show that the final maximum and minimum potentials for the surfaces are in the required range. Hence, the surfaces have been protected and the current at the anodes minimised.

Table 4: Initial and final currents applied to the anodes using the equation (18).

Name	Initial Current (mA)	Final Current (mA)
cpopt anode bow	−315.3	−482.0
cpopt anode keel	−300.1	−22.2
cpopt anode stern	−159.4	−6982.0

Table 5: Initial final currents applied to the anodes using the eqn (19).

Name	Initial Current (mA)	Final Current (mA)
cpopt anode bow	−315.3	−2616.0
cpopt anode keel	−300.1	−2524.2
cpopt anode stern	−159.4	−3339.4

Table 6: Maximum and minimum potentials at damaged areas for the initial anode current, before the osptimisation.

Name	Maximum Potential (mV)	Minimum Potential (mV)	Range
cpopt anode bow	−644.0	−645.3	×
cpopt anode keel	−642.7	−652.3	×
cpopt anode stern	−633.5	−642.1	×

Up to now, the Sequential Linear Programming (SLP) has been the method of optimisation used. Two more methods of optimisation were investigated, Sequential Quadratic Programming (SQP) and Modified method of Feasible Directions (MFD), and analysed below.

1.5.1 Sequential Quadratic Programming method (SQP)

The Sequential Quadratic Programming method (SQP) was also studied for the same initial values of current at the anodes and the same range of constraints.

The objective function value in this case was: 8285.8mA.

The current at the anodes after the optimisation:

Table 9 shows the final maximum and minimum potential at the surfaces for this method of optimisation. All the values are within the required protection range.

The objective function was not as much reduced as with the SLP. In addition, the anode at the keel still supplies a significant amount of current to the hull.

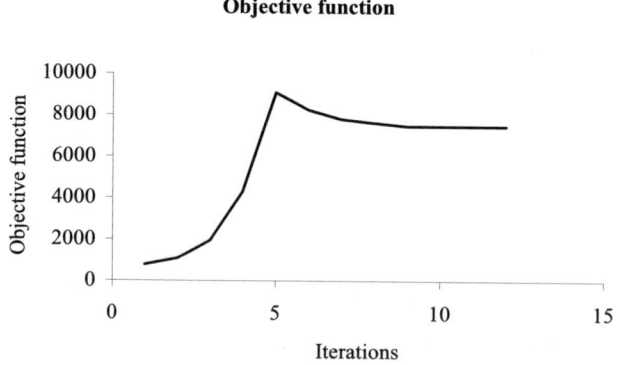

Figure 7: Evolution of the objective function.

Table 7: Maximum and minimum potentials at damaged areas after the optimisation.

Name	Maximum Potential (mV)	Minimum Potential (mV)	Range
cpopt anode bow	−850.0	−870.9	✓
cpopt anode keel	−869.2	−1067.6	✓
cpopt anode stern	−850.0	−968.8	✓

Table 8: Final currents applied to the anodes, SQP.

Name	Final Current (mA)
cpopt anode bow	−2167.3
cpopt anode keel	−2133.6
cpopt anode stern	−3984.9

1.5.2 Modified method of Feasible Directions (MFD)

The Modified method of Feasible Directions (MFD) was studied for the same initial values of current at the anodes and the same range of constraints.

The objective function value in this case was: 7478.3mA.

The current at the anodes after the optimisation:

Table 11 shows the final maximum and minimum potential at the surfaces for this optimisation method. All the values are within the required protection range.

The anode at the keel was removed from the design since it was considered not beneficial to the purpose of minimising the current supplied.

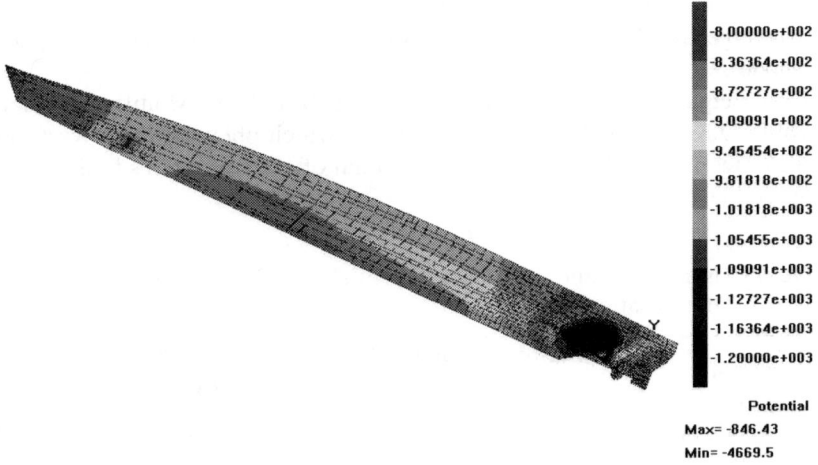

Figure 8: Potential distribution of the hull after the optimisation.

Table 9: Maximum and minimum potentials at the damaged areas after the optimisation, SQP.

Name	Maximum Potential (mV)	Minimum Potential (mV)	Range
cpopt anode bow	−907.9	−924.7	✓
cpopt anode keel	−924.3	−1014.9	✓
cpopt anode stern	−850.0	−957.3	✓

Table 10: Initial final currents applied to the anodes, MFD.

Name	Final Current (mA)
cpopt anode bow	−503.3
cpopt anode keel	0.0
cpopt anode stern	−6975.0

1.5.3 Comparison

In this experiment, the current at the anodes have been minimised, more precisely, the current at the anodes have been adjusted to supply the minimum amount that keeps the surfaces within the protection range.

The objective functions, eqn (18) and eqn (19) were compared, the first one reduced the amount of current required to achieve protection of the surfaces by reducing the utilisation of the anodes. Equation (19) has the affect of minimising the differences between the anode currents and maintaining them at a level sufficient to protect the vessel. Therefore, unlike the first case, the anodes will not be completely removed as the current is spread amongst the anodes.

The objective function, eqn (18), was used to minimise the current supplied by the anodes.

The Sequential Linear Programming (SLP) and the Modified method of Feasible Directions (MFD) are the methods which obtained the lowest values of the objective function and led the potential of all the surfaces to the required range (table 12).

Table 11: Maximum and minimum potentials at the damaged areas after the optimisation, MFD.

Name	Maximum Potential (mV)	Minimum Potential (mV)	Range
cpopt anode bow	−850.0	−870.8	✓
cpopt anode keel	−868.2	−1067.1	✓
cpopt anode stern	−850.0	−968.5	✓

Table 12: Comparison of the objective function value for the SLP, SQP and MFD.

	SLP	**SQP**	**MFD**
Objective Function (mA)	7486.3	8285.8	7478.3

The Modified method of Feasible Directions (MFD) was the method of optimisation which obtained the smallest value of the objective function, the SLP had a value 0.11% higher than the MFD and the SQP had still a value 10.80% higher (table 12).

The two methods which obtained the minimum values of the objective function, SLP and MFD, significantly reduced the current at the anodes at the bow and at the keel. In addition, the anode placed at the keel was considered unnecessary since the anode at the stern can supply most of the current to protect the damaged areas (table 13).

Finally, a study of the number of solutions evaluated by each one of the methods, in their search for the optimum, showed that the SQP was the method which required the fewest solutions (table 14). However, it did not find an appropriate value to the problem. The MFD found the best value, but it evaluated many solutions to obtain the results. The SLP found an adequate value in a reasonable number of solutions; consequently, it is considered the most suitable method.

1.5.4 Submarine

The *submarine model* (Section 1.4) was tested in order to validate the conclusions obtained from the previous model.

Table 13: Comparison of the final current at the anodes for the SLP, SQP and MFD.

Name	**Final Current (mA), SLP**	**Final Current (mA), SQP**	**Final Current (mA), MFD**
cpopt anode bow	−482.0	−2167.3	−503.3
cpopt anode keel	−22.2	−2133.6	0.0
cpopt anode stern	−6982.0	−3984.9	−6975.0

Table 14: Solutions evaluated by each one of the methods.

	SLP	**Solution Found**	**SQP**	**Solution Found**	**MFD**	**Solution Found**
Number of Solutions	42	✓	25	✗	96	✓

✓ -> Global Optimum Solution found
✗ -> Global Optimum Solution not found, local minimum

Table 15: Area of the anodes.

Position	Area(m^2)
Anode at the port	2.26
Anode at the starboard	2.26

Table 16: Area of the damaged areas.

Position	Area(m2)
Area at the bow	22.80
Area at the stern	121.40

Two anodes were placed on each side of the submarine hull in symmetric positions in order to protect the damaged areas (fig. 9). Their areas are shown in table 15.

The coating was considered slightly damaged at the areas shown in fig. 10. This was represented as an equivalent to the 50% of current of the steel polarisation curve, following coating thickness studies achieved by DiGiorgi [19, 20]. Their areas are shown in fig. 10. In addition, the propeller was set made of Nickel-Aluminium-Bronze with an area of 84.13m^2.

The current at the anodes was adjusted to supply the minimum amount which keeps the surfaces within the protection range by using three methods of optimisation.

The conclusions obtained with the frigate model are validated by this other model since the Sequential Linear Programming (SLP) and the Modified method of Feasible Directions (MFD) are also the methods which obtained the lowest values of the objective function and current at the anodes (table 18) and led the potential of all surfaces to the required range (table 17).

In addition, the SQP was proved once again to be the less accurate but the fastest method. The conclusions about the MFD and SLP are also applicable to this model. The MFD found the best solution, but it evaluated many solutions to obtain the results. The SLP found an adequate solution in a reasonable number of solutions; hence, it is considered the most suitable method.

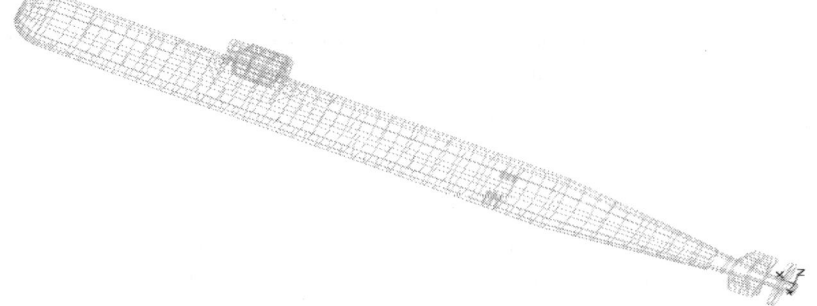

Figure 9: Anodes distribution on the submarine hull.

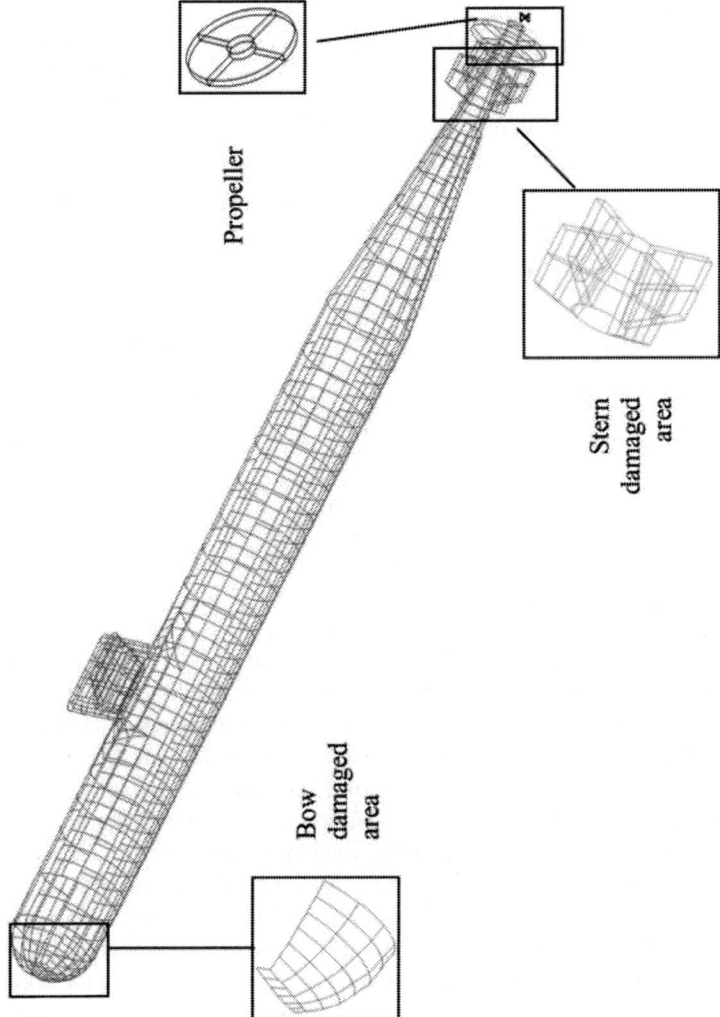

Figure 10: Propeller and damaged areas on the submarine hull.

Table 17: Comparison of the objective function value for the SLP, SQP and MFD, for the submarine model.

	SLP	SQP	MFD
Objective Function (mA)	59401.3	63073.1	59186.4

Table 18: Comparison of the final current at the anodes for the SLP, SQP and MFD, for the submarine model.

Name	Final Current (mA), SLP	Final Current (mA), SQP	Final Current (mA), MFD
cpopt anode starboard	−29701.5	−31536.7	−29593.1
cpopt anode port	−29699.7	−31536.4	−29593.3
Total	*−59401.2*	*−63073.1*	*−59186.4*

Table 19: Solutions evaluated by each one of the methods for the submarine model.

	SLP	Solution Found	SQP	Solution Found	MFD	Solution Found
Number of Solutions	13	✓	6	✓	26	✓

1.6 Achieving a uniform protection potential

Cathodic protection can reduce the roughness of the hull provided that a uniform potential distribution is supplied throughout the structure. Hull roughness has a dramatic effect on efficiency. Traditional thinking is that the smoother the hull, the better the performance. Research by numerous scientists has shown that there is a direct relation between the overall roughness of a hull, and the amount of energy required to propel it through water at a given speed. The rule-of-thumb in this regard is that for every 10 microns of increased roughness, there is a 1% increase on fuel requirement in the range of $0 - 250$ microns. Beyond 250 microns, the increase is 0.5% for every 10 microns. On most ships, the two factors causing drag will be the shape drag and the surface drag. The shape or form drag will be a factor which relates the speed of the vessel to the ship's length, see optimum design of a planning boat's hull from Yoshida (1997) [21], while the surface drag will relate the drag to the under water surface area of the vessel.

There are many factors that affect roughness. The first is the mechanical condition of the actual structure. Finally, and most importantly is the condition of the paint which is a major factor. As the paint covers the entirety of the structure, its roughness is the most important mechanical factor that affects the overall efficiency of the hull.

A British study, based on 400 ships, revealed that a new-built ship typically has a paint roughness that is about 125 microns. With use this roughness increases by 20 – 70 microns per year. The roughness of the hull starts increasing from the day the vessel is launched. The ships in the 20 micron increase range had excellent quality paints, plus cathodic protection, whereas the ships in the 70 micron range had traditional paints. Therefore, it can be seen that an annual increase of 2 – 7% in fuel costs is incurred as a result [22].

Corrosion can be produced on the surface of the structures when they are not within the required potential range. Earlier, the current at the anode was minimised to reduce the consumption keeping the surfaces within the range. Below, the current at the anodes are varied in such a way that the potential is matched to a target value. Thus, the surface will have a homogeneous potential value, which could avoid the roughness of the metal surfaces or detachment of the painting.

The least squares of the current potential on the elements on the surfaces with respect to a potential target value are used as objective function to achieve a uniform potential.

$$\text{Obj} = \sum_{i=1}^{n_s} \sum_{j=1}^{n_e} \left(\phi_{ij} - \phi_{\text{target},i} \right)^2 \tag{22}$$

Subject to the following constraints on the surface of the cathode:

$$g_i = \frac{\phi_i - \phi_{\text{Max},i}}{\left| \phi_{\text{min},i} \right|} \leq 0 \text{ on } \Gamma_C \quad i = 1,\ldots,m \tag{23}$$

$$g_j = \frac{\phi_{\text{min},j} - \phi_j}{\left| \phi_{\text{min},j} \right|} \leq 0 \text{ on } \Gamma_C \quad j = 1,\ldots,m \tag{24}$$

Where:

n_s is the number of surfaces.
n_e is the number of elements per surface.
ϕ_{target} is the target potential per surface.
n is the number of anodes.

Since constraints are considered, the final solution is guaranteed to be within the required potential range. The frigate model and the submarine model were analysed using three methods of optimisation.

1.7 Numerical examples, achieve a uniform protection potential

In order to ensure that the structure is protected, the potential of the damaged surfaces were required to be within a certain potential range –850mV and

−1100mV. In addition, in order to ensure the smoothness of the surface, a potential of −900mV was set as target value. The method of optimisation employed in this experiment was the Sequential Linear Programming (SLP) and it was applied on the frigate model (1.3).

The same damaged areas, physical properties and anodes location used to minimise the current at the anodes were employed in this experiment (fig. 4, table 1, fig. 5). The propeller was set made of Nickel-Aluminium-Bronze with an area of 1.84m².

The same initial currents were applied to each one of the anodes (table 3).

After the optimisation, the current at the anodes were increased to reach the potential range required (table 20).

The number of elements in the range of the potential required (−850mV, −1100mV) was zero at the beginning of the process since none of the damaged surfaces have their potential in the required range (table 6). At the end of the optimisation all the surfaces are within the required range (−850mV, −1100mV). In addition, the potentials were approached as much as possible to the target value −900mV.

During the optimisation process the number of elements in the range of the potentials required was increased until all the areas were protected. The objective function value is: 285996mV².

The potential distribution at the end of the optimisation is shown in fig. 11, where it can be seen that the distribution of potentials is more uniform around the target −900mV than in the contour plot obtained when the minimisation of the current at the anodes was carried out (fig. 8).

The Sequential Linear Programming (SLP) was the method of optimisation used. Two more methods of optimisation were analysed below, Sequential Quadratic Programming (SQP) and Modified method of Feasible Directions (MFD).

Table 20: Initial final currents applied to the anodes.

Name	Initial Current (mA)	Final Current (mA)
cpopt anode bow	−315.3	−1818.6
cpopt anode keel	−300.1	−1079.2
cpopt anode stern	−159.4	−5060.7

Table 21: Maximum and minimum potentials at the damaged areas for the initial anode current, before the optimisation.

Name	Maximum Potential (mV)	Minimum Potential (mV)	Range
cpopt anode bow	−890.3	−907.7	✓
cpopt anode keel	−910.2	−1033.6	✓
cpopt anode stern	−850.0	−960.1	✓

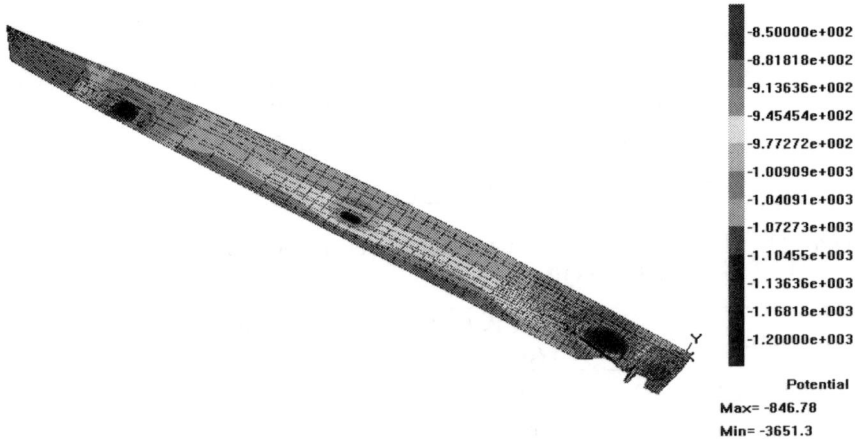

Figure 11: Potential distribution over the hull after the optimisation of the process of smoothing the potential, SLP.

1.7.1 Sequential Quadratic Programming method (SQP)

The Sequential Quadratic Programming method (SQP) was also studied for the same initial values of current at the anodes and constraints.

The objective function value in this case was: $1713417mV^2$.

The current at the anodes after the optimisation:

Table 23 shows that the final maximum and minimum potential at the surfaces were led to the required protection range. The distribution of potentials is shown in fig. 12, which has higher values than the one obtained with the SLP (fig. 11).

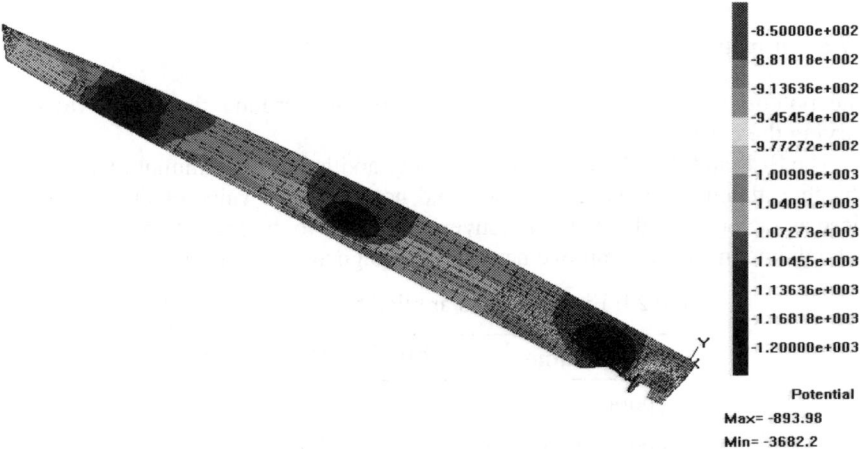

Figure 12: Potential distribution over the hull after the optimisation of the process of smoothing the potential, SQP.

Table 22: Initial final currents applied to the anodes, SQP.

Name	Final Current (mA)
cpopt anode bow	−2675.9
cpopt anode keel	−2646.8
cpopt anode stern	−4979.7

Table 23: Maximum and minimum potentials at the damaged areas after the optimisation, SQP.

Name	Maximum Potential (mV)	Minimum Potential (mV)	Range
cpopt anode bow	−962.3	−983.8	✓
cpopt anode keel	−983.6	−1099.4	✓
cpopt anode stern	−897.4	−1027.0	✓

1.7.2 Modified method of Feasible Directions (MFD)

The Modified method of Feasible directions (MFD) was studied for the same initial values of current at the anodes and the same value of the constraints.

The objective function value, in this case, was: $273074mV^2$.

The current at the anodes after the optimisation (shown in table 24):

Table 25 shows that the final maximum and minimum potential at the surfaces have been led to the protection range estimated. The distribution of potentials is shown in fig. 13, which has a similar range of values than the one obtained with the SLP (fig. 11). In addition, these results present a better distribution of the potential, discerning that the anode placed at the keel of the hull could be unnecessary for the purpose of keeping a uniform distribution of potential.

1.7.3 Comparison

The potential at the surfaces have been adjusted around the target value by varying the current at the anodes.

The SLP and the MFD behaved smoothly and led to the minimum value of the objective function. However, the SQP did not adjust the value of the potential as close as the other methods did. It converged rapidly to the protection range though, but it did not make any improvement of this design after this occurred (table 26).

Table 24: Final currents applied to the anodes, MFD.

Name	Final Current (mA)
cpopt anode bow	−2345.7
cpopt anode keel	−500.9
cpopt anode stern	−5078.9

Table 25: Maximum and minimum potentials at the damaged areas after the optimisation, MFD.

Name	Maximum Potential(mV)	Minimum Potential(mV)	Range
cpopt anode bow	−895.6	−910.9	✓
cpopt anode keel	−907.0	−1033.4	✓
cpopt anode stern	−850.0	−959.8	✓

Obviously, the current at the anodes obtained with the SQP were moderately high in comparison with the ones obtained with the SLP and MFD. The common factor after observing the table 27 is that the current at the anode at the stern is similar for the three methods after the optimisation since the propeller is placed nearby. The main difference between the SLP and the MFD is the distribution of the current at the anodes at the bow and at the keel. The MFD decided that the anode at the bow could supply more current to the damaged areas at the bow and at the keel and consequently reduce the amount of current supplied by the anode placed at the keel. This could suggest the idea of removing the anode at the keel in the final design.

In order to confirm this hypothesis, *the same model was optimised but only with two anodes*, one placed at the stern and the other one placed at the bow of the frigate. The final value of the objective function was 267404mV^2, which implies an improvement of 2% with the previous design. Table 28 also shows that a closer potential range is reached when only two anodes are employed.

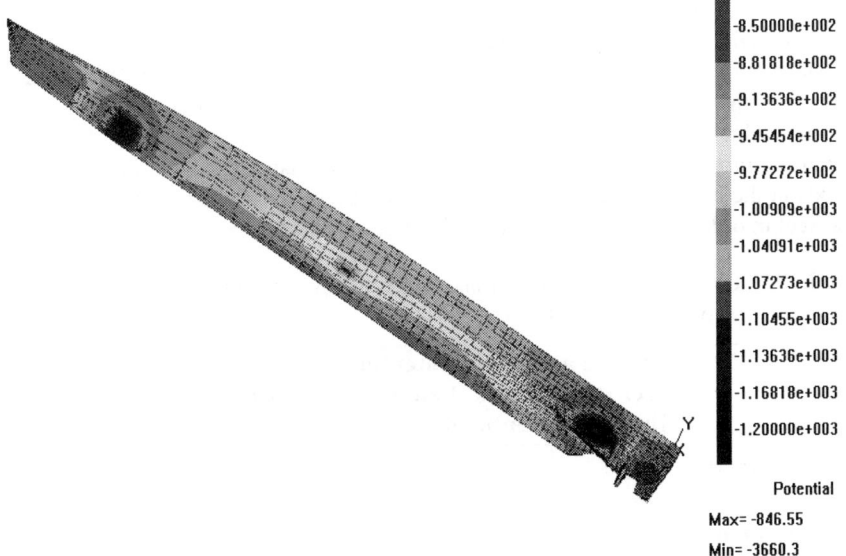

Figure 13: Potential distribution over the hull after the optimisation of the process of smoothing the potential, MFD.

Table 26: Comparison of the objective function value for the SLP, SQP and MFD.

	SLP	SQP	MFD
Objective Function (mV^2)	285996	1713417	273074

Table 27: Comparison of the final current at the anodes for the SLP, SQP and MFD.

Name	Final Current (mA), SLP	Final Current (mA), SQP	Final Current (mA), MFD
cpopt anode bow	−1818.6	−2675.9	−2345.7
cpopt anode keel	−1079.2	−2646.8	−500.9
cpopt anode stern	−5060.7	−4979.7	−5078.9

The study of the number of solutions evaluated by each one of the methods, in their search for the optimum solution, showed that the SQP was the method which required fewer solutions to find a feasible one but not the best. The MFD was the method which obtained the best solution but was also the one which required the most solutions. The SLP kept a compromise between accuracy and evaluated solutions (table 29).

The values of currents obtained when the current at the anodes were minimised and when the process of smoothing the potential was achieved were compared (table 30). Logically, the values obtained showed that the total current supplied by the anodes when their currents were minimised had lower values than when the potential was made uniform around the target value −900mV.

The main difference between both results is based on the current supplied by the anode placed at the bow. When the minimisation of the current at the anodes is achieved, the current supplied by the anode at the bow is reduced to a minimum amount which keeps the damaged area nearby within the constraints. In contrast, when the process of smoothing the potential on the damaged areas is achieved, the anode at the bow supplies a larger amount of current. This anode supplies current to the keel in order to keep the potential of its large surface as homogeneous as possible.

Table 28: Comparison between maximum and minimum values when 3 and 2 anodes were used.

	Achieve a uniform potential over the surfaces, 3 anodes, Min-Max potential difference	Achieve a uniform potential over the surfaces, 2 anodes, Min-Max potential difference
cpopt anode bow	15.3	15.6
cpopt anode keel	126.4	120.1
cpopt anode stern	109.8	108.6

Table 29: Solutions evaluated by each one of the methods.

SLP Solution Found		SQP Solution Found		MFD Solution Found	
25	✓	12	✓	42	✓

✓ -> Global Optimum Solution found
✓͞ -> Approximately the global optimum solution
✗ -> Global Optimum Solution not found, local minimum

Table 30: Comparison between the current at the anodes obtained when the minimisation of the current at the anodes and when the process of smoothing the potential over the surfaces were achieved.

	Minimise the current at the anodes, Final Current(mA), MFD	Achieve a uniform potential over the surfaces, Final Current(mA), MFD
cpopt anode bow	−503.3	−2345.7
cpopt anode keel	0.0	−500.9
cpopt anode stern	−6975.0	−5078.9
Total	−7478.3	−7925.5

Since both optimisations significantly reduce the value of the current at the anode at the keel, this anode could be considered unnecessary in the final design.

Table 31 shows the improvement in the difference between the maximum and minimum values of the potential on the surfaces obtained after the process of smoothing the potential in comparison with the difference obtained when the current at the anodes was minimised.

Table 31: Comparison of the potential difference between minimum and maximum values obtained on the damaged surfaces after minimising the current at the anodes and smoothing the potential over the surfaces.

	Minimise the currents of the anodes, Min-Max potential difference	Achieve a uniform potential over the surfaces, Min-Max potential difference
cpopt anode bow	20.8	15.3
cpopt anode keel	198.9	126.4
cpopt anode stern	118.5	109.8

Table 32: Comparison of the objective function value for the SLP, SQP and MFD, for the submarine model.

	SLP	SQP	MFD
Objective Function (mV²)	608254.0	913883.0	596490.0

Table 33: Comparison of the final current at the anodes for the SLP, SQP and MFD, for the submarine model.

Name	Final Current (mA), SLP	Final Current (mA), SQP	Final Current (mA), MFD
cpopt anode port	−29674.9	−31536.4	−29587.0
cpopt anode starboard	−29671.5	−31536.7	−29585.5

1.7.4 Submarine

The *submarine model* (1.4) was also tested in order to validate the conclusions obtained.

The same conclusions obtained with the frigate model are applicable to the submarine model. The SLP and the MFD behaved efficiently and led to the minimum value of the objective function but not the SQP (tables 32, 33).

Similar conclusions were obtained with respect to the solutions evaluated. The SQP was the method which required fewer solutions to find a feasible one but not the best. The MFD was the method which obtained the best solution computing just one solution more than the SLP (table 34).

2   Conclusions

In this work, a methodology has been developed to automate many of the processes necessary to design an ICCP system.

The modelling paradigm has been changed so that the corrosion engineer can define the objectives and requirements of the design for which the model will automatically find a solution. Replacing the typical modelling approach where the user defines all the conditions for which the model predicts the behaviour, thus requiring the user through a process of trial and error to iterate towards satisfying the design objectives.

Table 34: Solutions evaluated by each one of the methods for the submarine model.

	SLP	Solution Found	SQP	Solution Found	MFD	Solution Found
Number of Solutions	22	✓	8	✓	23	✓

The approach has been demonstrated on the ICCP system design of a frigate and submarine. The optimum anodes currents have been predicted to protect the vessels and a second design has been achieved which achieves a uniform protection potential on the surface.

The SLP, SQP and MFD were compared for the same model and for the two objectives studied. All the methods achieved final designs which substantially improved the initial solution, particularly the MFD and SLP. The Sequential Linear Programming (SLP) and the Modified method of Feasible Directions (MFD) were the methods which obtained the smallest values of the objective function and achieved protection potentials at all the damaged surfaces within the required.

The methodology can be applied to a wide range of corrosion applications including pipelines and oil and gas structures.

The approach has also been developed to apply the technology to the control of the associated corrosion related electric fields.

References

[1]   Adey, R.A., Brebbia, C.A., & Niku, S.M., Applications of Boundary Elements in corrosion engineering, Topics in BEM Research, Computational Mechanics Publications, 1990.

[2]   John Morgan. Cathodic Protection. National Association of Corrosion Engineers, NACE, 1987.

[3]   Adey, R.A., Niku, S.M., & Brebbia, C.A., Computer-aided-design of cathodic protection systems, et al, *Appl Ocean Res* **8**:(**4**) 209–222, October 1986.

[4]   Niku, S.M. & Adey, R.A., A CAD system for the analysis and design of cathodic protection systems, Institution of Corrosion Science and Technology, Chapter 13, *Plant Corrosion: Prediction of Materials Performance*, 1985.

[5]   Adey, R.A. & Hang, P.Y., Computer Simulation as an Aid to Corrosion Control and Reduction, Corrosion 99 Conference, San Antonio, Texas, USA, Nace, 1999.

[6]   Santana Diaz, E. & Adey, R., A Computational Environment for the Optimisation of CP system Performance and Signatures, Warship CP2001. Shrivingham. UK.

[7]   Adey, R.A. & Hang, P.Y., Optimum Design of Ship Corrosion Protection Using Computer Simulation, Computers and Ships Conference, The Institute of Marine Engineers, London, UK, May 1999.

[8]   Adey, R.A. & Baynham, J.M.W., Design and Optimisation of Cathodic Protection Systems Using Computer Simulation, NACE's Annual Conference "Corrosion 2000", Orlando, Florida, USA, March 2000.

[9]   Santana Diaz, E., Adey, R., & Baynham, J., Optimising the location of anodes in cathodic protection systems, OPTI 2003, May 2003.

[10]   Zamani, N.G. & Chuang, J.M., Optimal control of current in a cathodic protection system: A numerical investigation. *Optimal Control Applications & Methods*, Vol. 8, 339–350, 1987.

[11] Kishimoto, K., Amaya, K., Miyasaka, M., & Aoki, S., Boundary Element Analysis on Galvanic Corrosion. (computational accuracy and optimisation of cathodic protection). In Advances in boundary element method in Japan and USA. Computational Mechanics Publication, Southampton (*Topics in Engineering Vol. 7*), pp. 403–414, 1990.

[12] Kishimoto, K., Amaya, K., & Aoki, S., Optimization of cathodic protection using boundary element method. In *Boundary Element Methods. Principles and Applications*, ed. M. Tanaka and Q. Du. Pergamon Press, Oxford, pp. 329–338, 1990.

[13] Fontana, M.G. & Greene, N.D., *Corrosion Engineering*, 3rd Edition, McGraw-Hill, New York, 1986.

[14] Amaya, K. & Aoki, S., Optimum design of cathodic protection system by 3-D BEM, *Boundary Element Technology*, VII, pp. 375–388, 1992.

[15] Hou, L.S. & Sun, W., Numerical Methods for Optimal Control of Impressed Cathodic Protection Systems, *International Journal for Numerical Methods in Engineering*, Vol. **37**, 2779–2796, 1994.

[16] Aoki, S., Amaya, K., & Gouka, K., Optimal cathodic protection of ship, In *Boundary Element Technology XI*, ed. R.C. Ertekin, C.A. Brebbia, M. Tanaka & R. Shaw. pp. 345–356, 1996.

[17] Pei Yuan Hang & Adey, R.A., *Boundary Elements XXI*, eds. C.A. Brebbia, H. Power, pp. 195–206, 1999.

[18] Vanderplaats, G.N., DOT/DOC Users Manual, Vanderplaats, Miura and Associates, 1993.

[19] DeGiorgi, V.G. & Hamilton, C.P., Coating Integrity Effects on ICCP System Parameters, *Boundary Elements XVII*, Computational Mechanics Publications, 395–403, 1995.

[20] DeGiorgi, V.G., Finite resistivity and shipboard corrosion prevention system performance, *Boundary Elements XX*, Computational Mechanics Publications, 555–563, 1998.

[21] Yoshida, Y., Optimum Design of a Planing Boat's Hull Form, Marine Technology III, Editors: T. Graczyk, T. Jastrzebski, C.A. Brebbia, pp. 3–13, 1999.

[22] http://www.hydrex.be/MAG_files/Mag62-04.htm

Coating prediction from reference cells measurements and eventual assistance of the UEP

E.S. Diaz & R.A. Adey
C.M. BEASY Ltd, UK.

Abstract

Damages appear on a hull of a vessel during its lifetime. In many cases the correct position of these damages are completely unknown. Its knowledge is important from mainly two points of view.

- Cathodic protection of the vessel. A corroded hull would be economically inefficient (current and fuel consumption) apart from being potentially dangerous since it is a hot point for crack corrosion [1].
- Noisiness of the vessel. Damages will increase the current flux from the anode to the cathode (damaged areas). This will augment the noise of the vessel and therefore it will become easily detectable to the enemy [2–4].

The aim of this work is to detect the damage by using the information available from the ICCP system and/or the signature.

1 Introduction

A vessel can be detected from its surrounding magnetic fields. This is due to two main sources:

- The magnetic field associated with the permanent/induced magnetism, present because of the material used in the construction of a vessel and the earth's magnetic field.
- The electrical currents driven by the ship into the sea. The principal sources of these currents are related with ship corrosion or the ICCP system. These magnetic fields are named Corrosion Related Magnetic Fields (CRM). The CRM can be a significant proportion of a ship's magnetic signature for vessels constructed using non-magnetic materials [5].

Modelling techniques normally start from an assumed condition of the vessel. Given some assumed condition, the level of protection provided by the cathodic protection system can be predicted as well as the corrosion related electric and magnetic fields. Therefore, designers when assessing the signature and the effectiveness of the CP system will perform tests based on a number of possible conditions of the hull expected over the life cycle of the ship.

In typical CP systems, the designer knows the source of current (the anodes) but does not have a clear knowledge of where the current goes (the cathode) as this depends upon the condition of the metallic surfaces, *etc.* A method is therefore presented to determine where the current goes from the anodes and hence predict the general condition of the vessel and possible areas of damage. Once this information is known the associated electric and magnetic signatures can be predicted.

In addition, the detection of the areas of the vessel, which are acting as sinks of current, is of vital importance in order to know which part of the structure is disclosing the vessel. Their detection is a difficult matter that generally has to be solved in dry-docks by measuring the thickness of the coating with ultrasonic devices. However, to pull the ship out to the dry-dock to study the coating state is extremely costly. Moreover, on some occasions the area of the vessel which is taking current is concealed and cannot be easily detected even with the method indicated before.

In this chapter, the coating state of a structure, position and current taken by the damage, is analysed by using the information related with some sensors placed on the hull of the structure. The minimum amount of information necessary to carry out the prediction is searched for. Data will be presented showing the sensitivity of the predictions to the accuracy of the data and the number of reference cells, for example. Extra information of the Under Electric Potential (UEP) of the vessel will be included and their influence in the prediction will be shown.

Readers are advised that this paper can be viewed at the website <http://www.beasy.com/publications/paper/Predicting_Coating_Conditions.pdf> where the figures can be seen in their original colour format.

1.1 State of the art

In 1996, Aoki, Amaya and Gouka [6] applied the boundary element method to detect a paint defect on the hull of a ship. A painted hull with cathodic protection applied from some impressed anodes was studied, where the paint was assumed to be damaged during navigation. The damaged area could be accurately predicted with this method when the damage was located only on one element of the hull. However, the effectiveness of the method reduced significantly when the size of the damaged area was larger than an individual element. As the damaged size is unknown, the number of elements to use in the search is also unknown. A further complexity occurred where the damage had different coating thicknesses.

In a real case the unknowns are the following:

- The size of the damaged areas.
- The number of damaged areas.

– The position of the damaged areas.
– The coating thickness of the damaged areas.

This method could not cope with this amount of unknowns without carrying out a huge number of combinations.

In the approach proposed, an optimisation based search is performed to match the coating state of the surface. The predicted coating state is achieved by matching some potential reference data on the hull of the structure. Extra information, such as the Underwater Electric Potential (UEP) of the vessel, can also be included in the search. The quantity of information required, in particular the potential reference data, to obtain a reasonable solution is also studied.

1.2 Interpolation method

In order to identify the damaged areas, the coating of the vessel will be automatically modified by the optimisation method [7]. At least two polarisation curves should be considered, one with almost fully coated surface and the other with almost fully uncoated surface, representing the material underlying the coating. Once the value of the coating is found amongst the curves, a simple interpolation will provide the correct value of the current and potential.

The accuracy of the polarisation curve is an important factor in the simulation of a cathodic protection problem if accurate predictions are to be made of the protection potential and the current. However, in this case, an accurate polarisation curve of the underlying material was found not to be needed. This conclusion was obtained after using a linear polarisation curve to predict the state of the coating. A possible explanation for this unexpected result is that in the models solved, the ship hull was normally considered to be coated with a 'perfect paint' except in the areas where there was damage. Therefore the polarisation curve was simply used to differentiate between areas of the hull where the current flow into the hull was zero (the perfectly painted areas) and the areas where there was current flow into the hull (the damaged areas). Further studies would be necessary to test this hypothesis for cases where the general condition of the hull required different percentages of coating damages.

Nevertheless, successful results were obtained using realistic polarisation data as well.

The term coating sensors will be employed, from now on, as point positions on the surface of the hull at which the coating will be modified (fig. 6) by the optimisation surface (variables). The coating of the rest of the surface will be interpolated amongst these coating sensors. Two methods of interpolation were investigated, Radial Basis functions and three closest coating sensors.

1.2.1 Radial Basis function interpolation

Radial basis functions (RBFs) are a class of functions that exhibit radial symmetry, that is, they may be seen to depend only, apart from some known parameters, on the distance $r = \| x - x_j \|$ between the centre of the function and a generic point x. These functions can be generically represented in the form $\phi(r)$. This means that there exist infinite radial basis functions [8].

These functions may be classed into: globally supported and compactly supported ones depending on their supports, this is to say, whether they are defined on the whole domain or only on part of it.

Those most employed within the globally supported RBFs are:

$$\text{Multiquadratic(MQ)} \quad \sqrt{(x-x_j)^2 + c_j^2}, \ c_j > 0 \tag{1}$$

Reciprocal

$$\text{Multiquadratic (RMQ)} \quad \left((x-x_j)^2 + c_j^2\right)^{-\frac{1}{2}}, \ c_j > 0 \tag{2}$$

$$\text{Gaussians (G)} \quad \exp\left(-cr^2\right), c_j > 0 \tag{3}$$

$$\text{Thin-plate splines (TPS)} \quad r^{2\beta} \ln r, \beta \in M \tag{4}$$

Where:

c is a coefficient.
M is an integer number.

Within the compactly supported RBFs are:
Wu and Wendland,

$$(1-r)^t + q(r) \tag{5}$$

Where:

q(r) is a polynomial and $(1-r)^t$ is 0 for r greater than the support.
t is the polynomial order of $(1-r)^t$

Buhmann,

$$\frac{1}{3} + r^2 - \frac{4}{3}r^3 + 2r^2 \ln r \tag{6}$$

The previous RBFs were attempted for the problem of predicting the coating using coating sensors points. The best predictions were achieved by the equation shown below:

$$(1-r)^2 \tag{7}$$

Briefly, an interpolation with RBFs may take the form:

$$s(p) = \sum_{j=1}^{N} \alpha_j \varphi\left(\|p - p_j\|\right) \tag{8}$$

In this case:

$$s(p) = \sum_{j=1}^{N} \alpha_j (1-r)^2 = \sum_{j=1}^{N} \alpha_j \left(1 - \|p - p_j\|\right)^2 \tag{9}$$

Where:

N is the number of generic points.
p is the generic point.

The values of s(p) are known, coating values, and therefore the set of equations of the form:

$$s(p_1) = \sum_{j=1}^{N} \alpha_j \left(1 - \|p_1 - p_j\|\right)^2 \tag{10}$$

$$s(p_2) = \sum_{j=1}^{N} \alpha_j \left(1 - \|p_2 - p_j\|\right)^2 \tag{11}$$

$$s(p_N) = \sum_{j=1}^{N} \alpha_j \left(1 - \|p_N - p_j\|\right)^2 \tag{12}$$

The α_j parameters are obtained by solving the above system of equations.

Once the α_j are found, eqn (9) can be applied to all the points of the surface.

The generic points are our coating sensors points, which are the variables of the optimisation.

1.2.2 Three closest coating sensors

The three closest coating sensors can be employed to compute the coating of the current point by using a linear interpolation.

The process is based on the next steps:

1. The three closest coating sensors to the point considered, *i*, are searched for; their coating values, $s(p_1)$, $s(p_2)$, $s(p_3)$, and their distances to the considered point, d_{i1}, d_{i2}, d_{i3}, are taken.
2. The equivalent distance (d_{eq}) is computed.

$$\frac{1}{d_{eq}} = \frac{1}{d_{i1}} + \frac{1}{d_{i2}} + \frac{1}{d_{i3}} \tag{13}$$

3. And the value of the coating at the point considered is computed:

$$s(p_i) = \frac{d_{eq}}{d_{i1}} s(p_1) + \frac{d_{eq}}{d_{i2}} s(p_2) + \frac{d_{eq}}{d_{i3}} s(p_3) \tag{14}$$

The eqns (9) and (14) were used to predict the coating of the surface, Radial Basis Functions behaved slightly better than the three closest coating sensors. Thus, RBFs was the method of interpolation selected to implement the experiments below. Some additional experiments are implemented with the three closest coating sensors at the end of this chapter.

1.3 Prediction of the coating from reference cells measurements, objective function and constraints

The optimisation process requires the problem posed in the form of an objective function, design variables and constraints. The objective function was defined as the sum of the squares of the difference between the target potentials at the reference cells and the potentials predicted by the model (15) in order to match the measured reference cell potentials.

$$\text{Obj}= \sum_{i=1}^{i=n} \left(V_{t_i} - V_i\right)^2 \qquad (15)$$

Subject to the following constraints on the surface of the cathode:

$$g_i= \frac{V_i - V_{max,i}}{\left|V_{min,i}\right|} \leq 0 \text{ on } \Gamma_C \quad i= 1,\ldots,m \qquad (16)$$

$$g_j= \frac{V_{min,j} - V_j}{\left|V_{min,j}\right|} \leq 0 \text{ on } \Gamma_C \quad j= 1,\ldots,m \qquad (17)$$

Where:

        n is the number of references cells potential.
        V_t is the target potential at the reference cell.
        V_{min} is the minimum potential required at a reference cell.
        V_{max} is the maximum potential required at a reference cell.
        V is the computed potential at a reference cell.
        Γ_C is the surface of the cathode.
        m is the number of elements on the surface.

Note: It is normally recommended for the optimisation algorithms [9] that to obtain a well-conditioned problem the constraints will have roughly the same order of magnitude as the gradients (within a factor of 100). Therefore, the denominator of eqns (16) and (17) were applied to scale the constraints into non-dimensional units.

    The constraints were applied to limit the search space to that of practical significance with the resulting benefit that the speed of the solution was improved.

The method of optimisation used was the Sequential Linear Programming (SLP) which is a *Multivariable search* method [9]. This procedure uses algorithms which are based on geometric or logical concepts to move rapidly from a starting point away from the optimum to a point near the optimum. In addition, they attempt to satisfy the constraints associated with the problem and the Kuhn-Tucker conditions [10] as they generate improved values of the model.

1.4 The models

Two models were used to check whether the prediction of the coating was accurate, a cylinder model and a frigate model.

1.4.1 Cylinder model, description
A model of a cylinder was considered to study the prediction of the coating state.
The dimensions of the cylinder are:
  - Length: 34.0m.
  - Diameter: 10.0m.
The model has the characteristics shown in fig. 1. The surface of the cylinder was considered fully coated, thus, no current leaked through its surface apart from the damaged areas studied in the experiments. The surface near one of the edges of the cylinder was set of being made of Nickel-Aluminium-Bronze.

The electrolyte considered was seawater with a resistivity of 20ohm·cms [11], that implies a conductivity of about 5S/m.

To speed up the solution and since the model is symmetric, only half of it was modelled. The model has 619 elements, including the surrounding box which simulates the electrolyte. The cylinder itself was modelled with 600 elements.

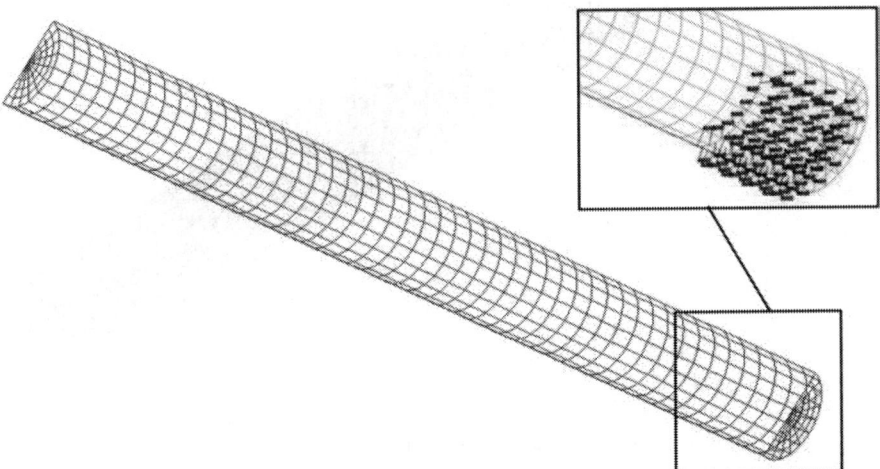

Figure 1: Cylinder model with Nickel-Aluminium-Bronze area. (Coloured versions of figure 1 and subsequent figures can be viewed at http://www.beasy.com/publications/papers/Predicting_Coating_Conditions.pdf).

1.4.2 Frigate model, description

A model of a frigate model was also considered to study the prediction of the coating state. The model has the characteristics shown in fig. 2. The frigate was considered fully coated apart from some specific damaged areas. The propeller was set made of nickel-aluminium-bronze.

To speed up the solution, and since the model is symmetric, only half of it was modelled.

The electrolyte considered was seawater with a resistivity of 20ohm·cms [11], that implies a conductivity of about 5S/m.

From the modelling perspective, the model has 1738 elements, including the surrounding box which simulates the electrolyte. The frigate itself was modelled with 1338 elements in most of the experiments analysed. A more refined mesh was created at the stern of the vessel since it is the most critical area of the frigate due to the propeller and location of the main anodes.

1.5 Cylinder model

Several sets of coating sensors were placed on the cylinder surface model (1.4.1) to study the effectiveness of the method. The potential data of the reference cells are to be added one by one to the optimisation to study their influence in the prediction of the coating state.

Figure 2: Frigate model. The dimensions of the frigate are:

- Waterline length: 34.0m.
- Draft: 2.3m.
- Waterline beam: 6.4m.

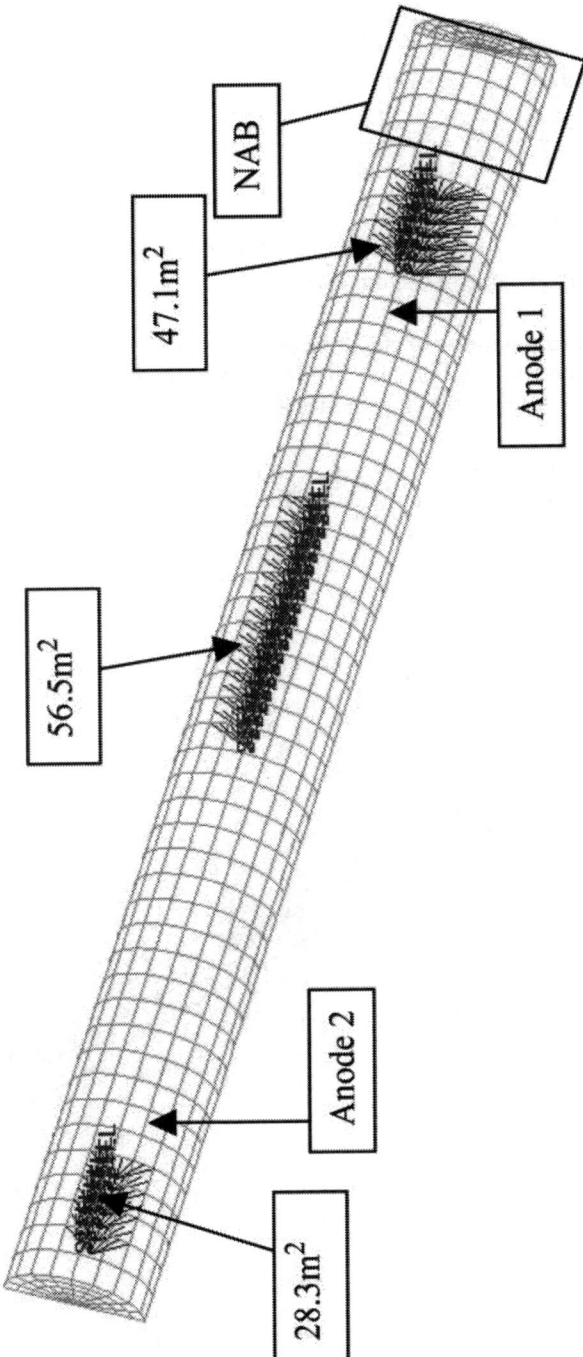

Figure 3: Damaged areas placed on the cylinder surface.

Table 1: Current at the anodes.

	Current Density (mA/m²)	Current (mA)
Anode 1	−6903.0	−21687.4
Anode 2	−4455.0	−13995.8

1.5.1 Damaged areas
Three damaged areas, bare steel, were placed on the cylinder surface. Figure 3 shows the position and size of the damaged areas.

1.5.2 Currents and positions of the anodes
Two impressed anodes were placed on the surface of the model (fig. 3). The currents supplied by each one of the anodes are shown in the table 1.

1.5.3 Coating search area
Figure 4 shows the search area in which the damaged areas are to be predicted.

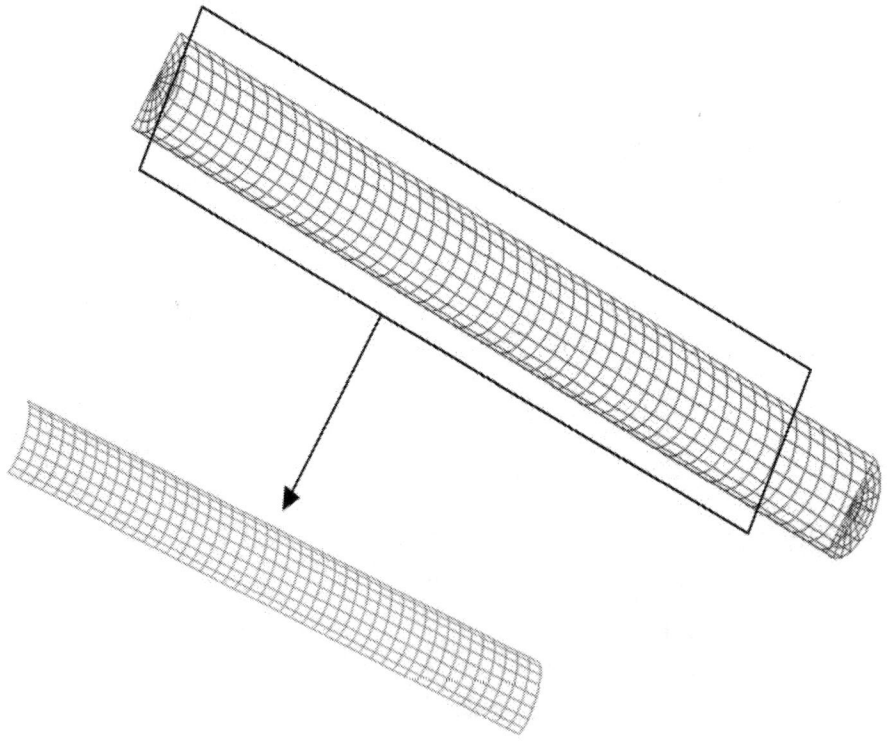

Figure 4: Search area on the cylinder structure.

Table 2: Target and constraints for five reference cells.

	Target (mV)	Constraint (mV)
REF1	−802.5	−770, −777
REF2	−788.5	−765, −775
REF3	−779.2	−755, −762
REF4	−828.5	−864, −870
REF5	−820.1	−815, −820

1.5.4 Reference cells

The number of reference cells, target values, was increased to determine how much data was required to detect the damage.

Figure 5 shows the position of the reference cells on the cylinder surface.

Tight constraints were set around the target values to make the optimisation software reach the target with more accuracy. The constraint values applied are shown in table 2.

1.5.5 First array of coating sensors, 7 coating sensors

An array of 7 coating sensors was placed on the predicting surface, distributed as it is shown in fig. 6. A radial basis function was employed as interpolation function to emulate the real state of the surface.

1.5.5.1 One reference cell The first reference cell potential was used as target value on the optimisation process (fig. 5).

Table 3 shows that only one damaged area at the edge of the search surface was found. There is not enough information to represent the real state of the surface since the constraints, despite being reasonably tight, are satisfied and the objective function presents a small value.

1.5.5.2 Two reference cells The second reference cell potential is added as target value on the optimisation process (fig. 5).

Table 4 shows that two damaged areas at the edges of the search surface were found. There is not enough information to represent the real state of the surface since the constraints, despite being reasonably tight, are satisfied. The objective function presents a higher value since the optimisation algorithm finds it difficult to match the sum of least squares and satisfy the constraints.

1.5.5.3 Three reference cells The third reference cell potential is added as target value on the optimisation process (fig. 5).

Table 5 shows that still only two of the three damaged areas are revealed. Again there is not enough information to represent the real state of the surface since while the constraints are satisfied, the objective function gives a high value.

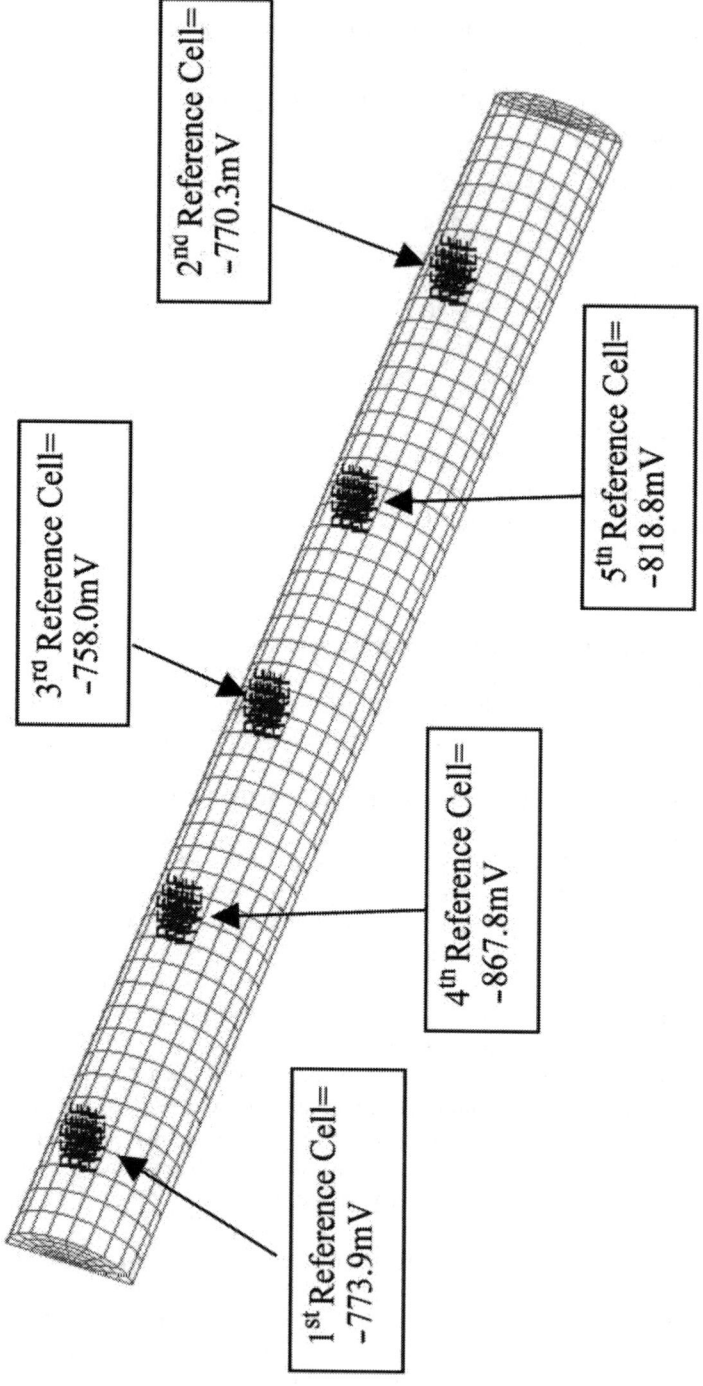

Figure 5: Position of the five reference cells on the cylinder surface.

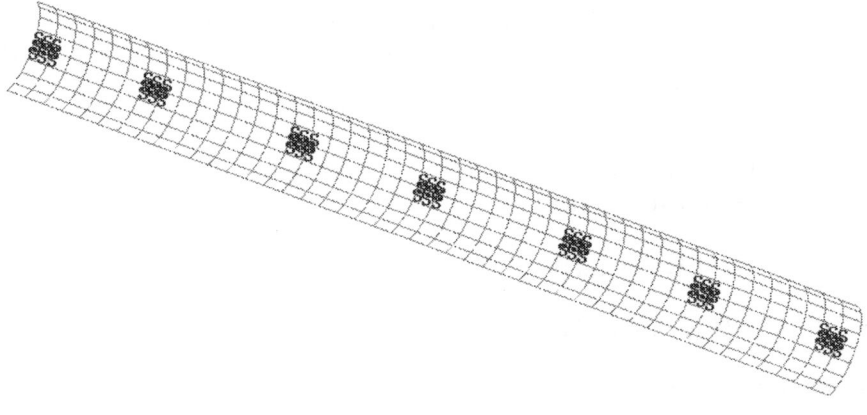

Figure 6: Seven coating sensors distribution on the prediction surface.

Table 3: Summary of results obtained when one reference cell was used as target value and 7 coating sensors to emulate the coating.

One reference cell	
	7 Coating sensors
Iterations	10
Objective function (mV²)	0.038
Number of references in the constraints	1/1
SURFACE REF1: **Potential (mV)**	−773.7
Absolute error target, 1ˢᵗ Ref Cell (mV)	0.2
Display	

Table 4: Summary of results obtained when two reference cells were used as target value and 7 coating sensors to emulate the coating.

Two reference cells

	7 Coating sensors
Iterations	9
Objective function (mV^2)	4.779
Number of references in the constraints	2/2
SURFACE REF1: Potential (mV)	−773.6
SURFACE REF2: Potential (mV)	−768.1
Absolute error target, 1st Ref Cell (mV)	0.3
Absolute error target, 2nd Ref Cell (mV)	2.2
Display	

1.5.5.4 Four reference cells The fourth reference cell potential is added as target value on the optimisation process (fig. 5).

In this case, table 6 shows that three damaged areas were found in an approximate correct position (fig. 3). The constraints are not satisfied due to the tight constraints around the target values. However, the optimisation tried to match them. Consequently, the final solution is the best solution which minimises the objective function and approaches the constraints.

1.5.5.5 Five reference cells The fifth reference cell potential is added as target value on the optimisation process (fig. 5).

In this case, table 7 shows that three damaged areas were also found in an approximate correct position (fig. 3). As previously, the constraints are not satisfied due to the tight constraints around the target values. However, the optimisation tried to match them. Consequently, the final solution is the best solution that minimises the objective function and approaches the constraints.

Table 5: Summary of results obtained when three reference cells were used as target value and 7 coating sensors to emulate the coating.

Three reference cells	
	7 Coating sensors
Iterations	6
Objective function (mV^2)	47.6
Number of references in the constraints	2/3
SURFACE REF1: Potential (mV)	−771.8
SURFACE REF2: Potential (mV)	−765.1
SURFACE REF3: Potential (mV)	−762.0
Absolute error target, 1st Ref Cell (mV)	2.1
Absolute error target, 2nd Ref Cell (mV)	5.2
Absolute error target, 3rd Ref Cell (mV)	4.0
Display	

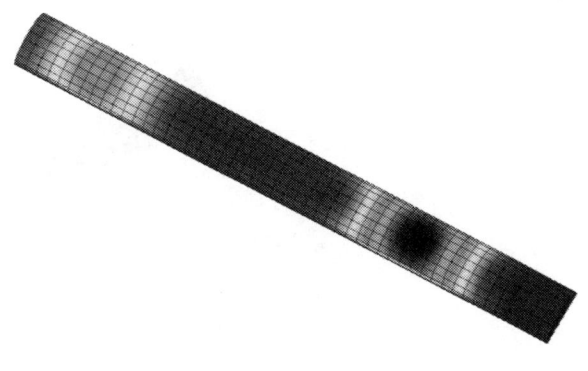

Two more sets of coating sensors (9 coating sensors and 20 coating sensors) where studied obtaining similar results to this first set of coating sensors.

1.5.6 Tests with 9 and 20 Coating sensors
A second and third array of 9 and 20 coating sensors were placed on the model (fig. 7). A set of tests were performed similar to that described earlier where the number of reference cell measurements were increased until the damage pattern was identified.

Table 6: Summary of results obtained when four reference cells were used as
target value and 7 coating sensors to emulate the coating.

Four reference cells	
	7 Coating sensors
Iterations	4
Objective function (mV^2)	3171.2
Number of references in the constraints	0/4
SURFACE REF1: Potential (mV)	−802.0
SURFACE REF2: Potential (mV)	−787.4
SURFACE REF3: Potential (mV)	−772.0
SURFACE REF4: Potential (mV)	−824.3
Absolute error target, 1st Ref Cell (mV)	28.1
Absolute error target, 2nd Ref Cell (mV)	17.1
Absolute error target, 3rd Ref Cell (mV)	14.0
Absolute error target, 4th Ref Cell (mV)	43.5
Display	

1.5.6.1 One reference cell The first reference cell potential is used as target
value (fig. 5)

1.5.6.2 Two reference cells The second reference cell potential is added as
target (fig. 5).

1.5.6.3 Three reference cells The third reference cell potential is added as target
value (fig. 5).

Table 7: Summary of results obtained when five reference cells were used as target values and 7 coating sensors to emulate the coating.

Five reference cells

	7 Coating sensors
Iterations	4
Objective function (mV^2)	3171.2
Number of references in the constraints	0/4
SURFACE REF1: Potential (mV)	−802.5
SURFACE REF2: Potential (mV)	−788.5
SURFACE REF3: Potential (mV)	−779.2
SURFACE REF4: Potential (mV)	−828.5
SURFACE REF5: Potential (mV)	−820.1
Absolute error target, 1st Ref Cell (mV)	28.6
Absolute error target, 2nd Ref Cell (mV)	18.2
Absolute error target, 3rd Ref Cell (mV)	21.2
Absolute error target, 4th Ref Cell (mV)	39.3
Absolute error target, 5th Ref Cell (mV)	1.3
Display	

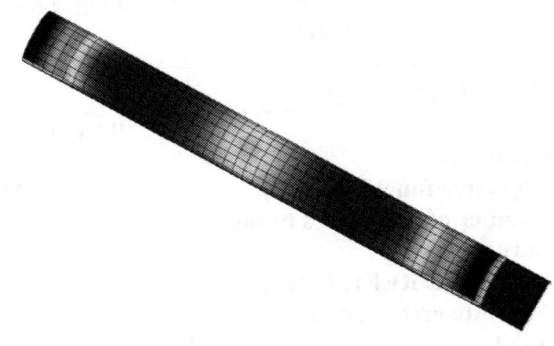

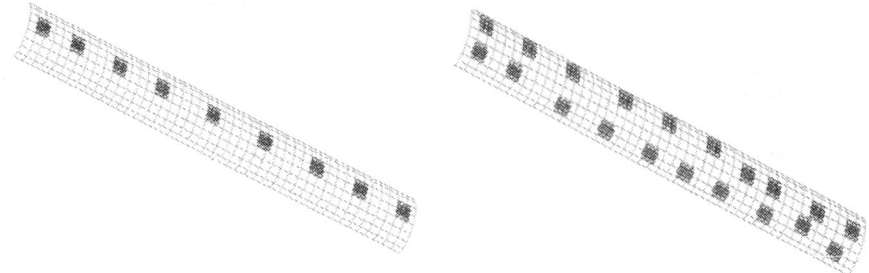

Figure 7: Nine and twenty coating sensors distribution on the prediction surface.

1.5.6.4 Four reference cells The fourth reference cell potential is added as target value (fig. 5).

1.5.6.5 Five reference cells The fifth reference cell potential is added as target value (fig. 5).

1.5.7 Summary, cylinder

The objective function increases its value with the number of reference cells, since the distribution of the coating sensors and the RBFs do not exactly represent the real situation on the damaged areas. Likewise, the constraint tolerance is not satisfied when the number of reference cells exceeds a certain number. The best results, in the experiment analysed, are obtained when the number of reference cells data is at least 4, regardless of the number of coating sensors. At that point, despite the constraint tolerance being exceeded, the results highlight the damaged areas and predicted the correct current flow in the model. The possible explanation for the good results being obtained is that the final design/solution is the best that could be achieved with the data provided. In general all the cases indicate that the more information available from the reference cells the better the prediction (table 13).

Table 8: Summary of results obtained when only one reference cell was used as target value with 9 and 20 coating sensors, respectively, to emulate the coating distribution.

One reference cell		
	9 Coating sensors	**20 Coating sensors**
Iterations	11	16
Objective function (mV^2)	0.81	0.06
Number of references in the constraints	1/1	1/1
SURFACE REF1: Potential (mV)	−773.0	−774.1
Absolute error target, 1st Ref Cell (mV)	0.9	0.2

Table 9: Prediction obtained when only one reference cell was used as target value with 9 and 20 coating sensors, respectively, to emulate the coating distribution.

	9 Coating sensors	**20 Coating sensors**
Display		

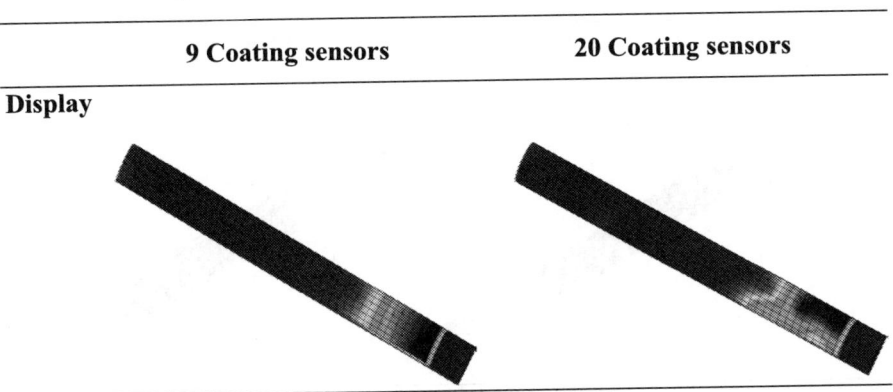

The constraints are satisfied when the number of reference cells is three or less but the solution does not completely predict all the damaged areas. However, when the number of reference cells is four or above the solution is clear but the constraints are not satisfied.

1.6 Frigate model

The frigate model was also considered to study the prediction of the coating state.

Two impressed anodes were placed on the surface of the model (Figure 8). The currents supplied by each one of the anodes are shown in the table 18.

Table 10: Summary of results obtained when two reference cells were used as target value with 9 and 20 coating sensors, respectively, to emulate the coating distribution.

Two reference cells	9 Coating sensors	20 Coating sensors
Iterations	11	11
Objective function (mV^2)	8.3	0.4
Number of references in the constraints	2/2	2/2
SURFACE REF1: Potential (mV)	−771.4	−773.7
SURFACE REF2: Potential (mV)	−768.8	−769.7
Absolute error target, 1st Ref Cell (mV)	2.5	0.2
Absolute error target, 2nd Ref Cell (mV)	1.5	0.6

Table 11: Prediction obtained when two reference cells were used as target value with 9 and 20 coating sensors, respectively, to emulate the coating distribution.

Display	9 Coating sensors	20 Coating sensors

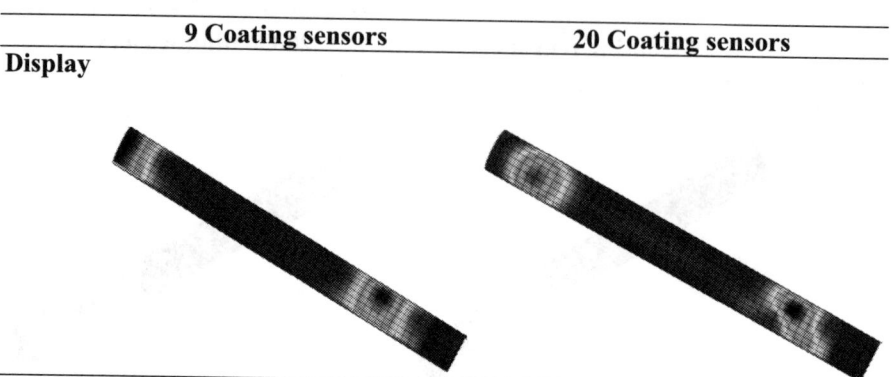

In order to test the methodology a test model was created by creating three areas of damage on the surface of the frigate, the model was solved and the results used to create the target data. Note that additional tests were performed to assess the impact of noise in the target data and other geometries. The results presented below are representative of those results.

The optimisation attempted to match the target model. An exact match will not be obtained because the damage was defined over discrete elements of the

Table 12: Summary of results obtained when three reference cells were used as target value with 9 and 20 coating sensors, respectively, to emulate the coating distribution.

Three reference cells

	9 Coating sensors	20 Coating sensors
Iterations	19	16
Objective function (mV^2)	9.5	0.7
Number of references in the constraints	3/3	3/3
SURFACE REF1: Potential (mV)	−776.4	−773.9
SURFACE REF2: Potential (mV)	−772.0	−770.4
SURFACE REF3: Potential (mV)	−757.3	−757.2
Absolute error target, 1st Ref Cell (mV)	2.5	0.0
Absolute error target, 2nd Ref Cell (mV)	1.7	0.1
Absolute error target, 3rd Ref Cell (mV)	0.7	0.8

Table 13: Prediction obtained when three reference cells were used as target value with 9 and 20 coating sensors, respectively, to emulate the coating distribution.

9 Coating sensors	20 Coating sensors
Display	

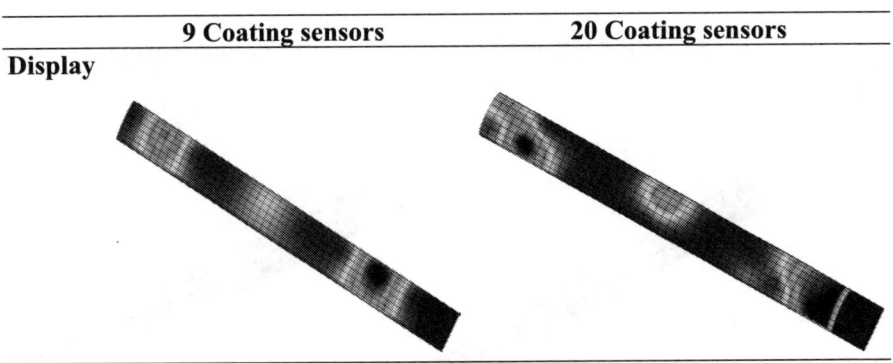

model whereas the coating sensors can only provide a general idea on the size and location of the damage.

Several sets of coating sensors were placed on the surface of the frigate model. Reference cells are to be included one by one in the process, as target values, to study their influence in the prediction of the coating.

Table 14: Summary of results obtained when four reference cells were used as target value with 9 and 20 coating sensors, respectively, to emulate the coating distribution.

Four reference cells		
	9 Coating sensors	**20 Coating sensors**
Iterations	4	17
Objective function (mV^2)	5253.4	2424.2
Number of references in the constraints	0/4	0/4
SURFACE REF1: Potential (mV)	−800.3	−797.3
SURFACE REF2: Potential (mV)	−830.1	−792.2
SURFACE REF3: Potential (mV)	−778.5	−777.0
SURFACE REF4: Potential (mV)	−844.1	−835.6
Absolute error target, 1st Ref Cell (mV)	26.4	23.4
Absolute error target, 2nd Ref Cell (mV)	59.8	21.9
Absolute error target, 3rd Ref Cell (mV)	20.5	19.0
Absolute error target, 4th Ref Cell (mV)	23.7	32.2

Table 15: Prediction obtained when four reference cells were used as target value with 9 and 20 coating sensors, respectively, to emulate the coating distribution.

	9 Coating sensors	20 Coating sensors
Display		

Table 16: Summary of results obtained when five reference cells were used as target value with 9 and 20 coating sensors, respectively, to emulate the coating distribution.

Five reference cells

	9 Coating sensors	20 Coating sensors
Iterations	4	14
Objective function (mV^2)	6339.1	2538.7
Number of references in the constraints	0/5	1/5
SURFACE REF1: Potential (mV)	−804.1	−803.6
SURFACE REF2: Potential (mV)	−831.5	−779.1
SURFACE REF3: Potential (mV)	−790.4	−774.9
SURFACE REF4: Potential (mV)	−851.0	−831.8
SURFACE REF5: Potential (mV)	−837.3	−818.2
Absolute error target, 1st Ref Cell (mV)	30.2	29.7
Absolute error target, 2nd Ref Cell (mV)	61.2	8.8
Absolute error target, 3rd Ref Cell (mV)	19.0	16.9
Absolute error target, 4th Ref Cell (mV)	32.2	36.0
Absolute error target, 5th Ref Cell (mV)	5.5	0.6

Table 17: Prediction obtained when five reference cells were used as target value with 9 and 20 coating sensors, respectively, to emulate the coating distribution.

	9 Coating sensors	**20 Coating sensors**
Display		

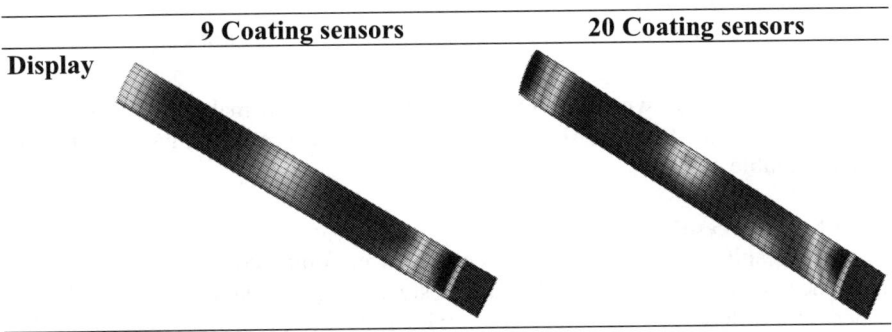

1.6.1 Damaged areas
Three damaged areas, bare steel, were placed on the surface of the frigate. Figure 8 shows the position and size of the damaged areas.

1.6.2 Coating sensors arrays
Two arrays of coating sensors were utilised in the frigate model, an array of 7 coating sensors and an array of 12 coating sensors distributed as shown in fig. 9.

A radial basis function was used to interpolate the data from the coating sensors on the hull surface.

1.6.3 Reference cells
The number of reference cells, target values, was increased to determine how much data was required to detect the damage.

Figure 10 shows the position of the reference cells on the hull surface.

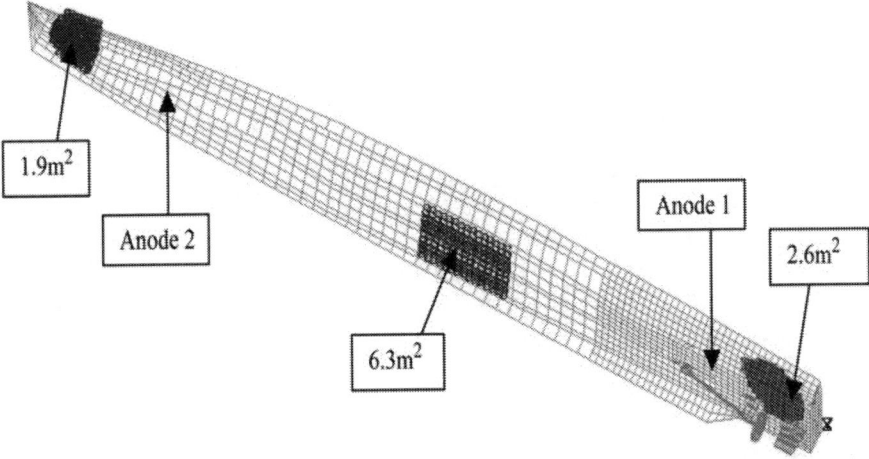

Figure 8: Damaged areas placed on the surface of the frigate and anodes position.

Table 18: Current at the anodes.

	Current Density (mA/m^2)	Current (mA)
Anode 1	−33554.1	−1813.7
Anode 2	−18262.1	−1649.9

Tight constraints were set around the target value to make the optimisation software reach the target with more accuracy. The constraint values applied are shown in table 19.

1.6.4 Frigate, results

The best results, in the experiment analysed, were obtained when the number of reference cells was at least 4. The final design clearly showed the damaged areas when four reference cells were used. The final design was still accurate but started to drift out of the range of constraints when the number of reference cells was above three.

The tables (table 20, table 21, table 22, table 23 and table 24) show how far the potential of each one of the surfaces studied are to the target one. The constraints are satisfied when the number of reference cells, target values, is three, two and one, but the solution is not complete. Some damaged areas are shown on the frigate hull, but other damaged areas are not shown yet since the information is insufficient.

When the number of reference cells is four or above the solution is clear. Increasing the number of coating sensors resulted in a reduction in the difference between the target potential and the final design potential.

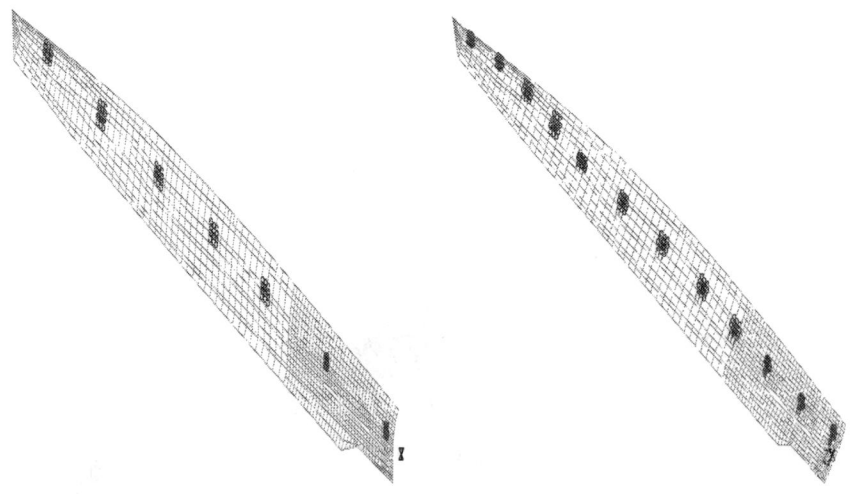

Figure 9: Seven and twelve coating sensors distribution on the surface of the frigate.

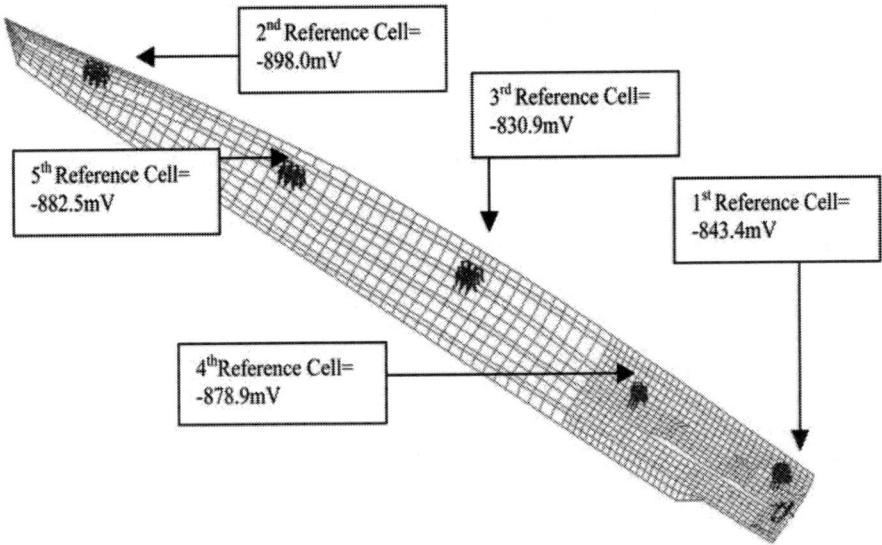

Figure 10: Position of the five reference cells on the hull surface of the frigate.

In the case with five reference cells, the solution obtained when the number of coating sensors was seven seems to be a better prediction than when the number of coating sensors was twelve. A possible explanation for this phenomenon is that the sensors in this case are placed closer to the bow of the frigate (fig. 9). Therefore, they are better able to represent the damage in that area. This assumption was confirmed with a new analysis carried out using only ten sensors (fig. 11 and fig. 12), removing the first and third sensors of the twelve previously had.

1.6.5 Summary, frigate
A frigate model has been analysed with three damaged areas on its surface. A compact radial basis function was used to interpolate the coating condition among the 'coating sensors'.

Table 19: Target and constraints for five reference cells.

	Target (mV)	Constraint (mV)
REF1	−843.4	−840, −850
REF2	−898.0	−895, −905
REF3	−830.9	−825, −835
REF4	−878.9	−873, −883
REF5	−882.5	−877, −887

Table 20: Results obtained using the 1st reference cell as target value and two different sets of arrays.

One reference cell		
	7 Coating sensors	**12 Coating sensors**
Iterations	13	11
Objective function (mV^2)	0.127	0.122
Number of references in the constraints	1/1	1/1
SURFACE REF1: Potential (mV)	−843.0	−843.7
Absolute error target, 1st Ref Cell (mV)	0.4	0.3
Display		

The number and position of reference cells, to determine where the damaged areas are, is not known in advance. In addition, a ship's ICCP system will be designed to provide the best ICCP system performance not necessarily to identify areas of damage. Consequently, if a reference cell is near a damaged area, the damage will be quickly predicted. However, reference cells placed away from the damaged areas will not give significant information unless the cells can provide an overall pattern sufficient to detect the damage. Therefore, in some cases the information provided is enough to predict, with accuracy, what the condition of the hull is, but in other occasions more information is needed. To avoid the uncertainty of this lack of information, several positions of the coating sensors should be studied. If the damaged areas appear consistently in the same area, then the damaged areas have been found.

Table 21: Results obtained using the 1st and the 2nd reference cell as target values and two different sets of arrays.

Two reference cells

	7 Coating sensors	12 Coating sensors
Iterations	9	9
Objgective function (mV^2)	2.368	0.755
Number of references in the constraints	2/2	2/2
SURFACE REF1: Potential (mV)	−842.3	−843.1
SURFACE REF2: Potential (mV)	−897.1	−897.2
Absolute error target, 1st Ref Cell (mV)	1.10	0.30
Absolute error target, 2nd Ref Cell (mV)	0.90	0.80
Display		

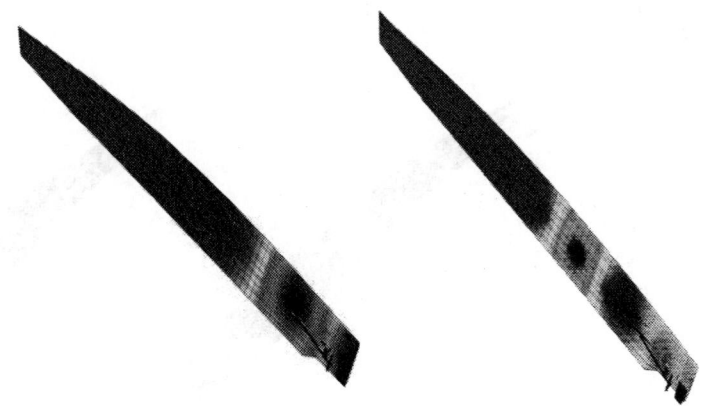

Table 22: Results obtained using the 1st, the 2nd and the 3rd references cell as target values and two different sets of arrays.

Three reference cells		
	7 Coating sensors	**12 Coating sensors**
Iterations	7	11
Objective function (mV^2)	40.255	2.79
Number of references in the constraints	3/3	3/3
SURFACE REF1: Potential (mV)	−843.1	−844.5
SURFACE REF2: Potential (mV)	−901.1	−898.5
SURFACE REF3: Potential (mV)	−825.4	−829.8
Absolute error target, 1st Ref Cell (mV)	0.30	1.10
Absolute error target, 2nd Ref Cell (mV)	3.10	0.50
Absolute error target, 3rd Ref Cell (mV)	5.50	1.10
Display		

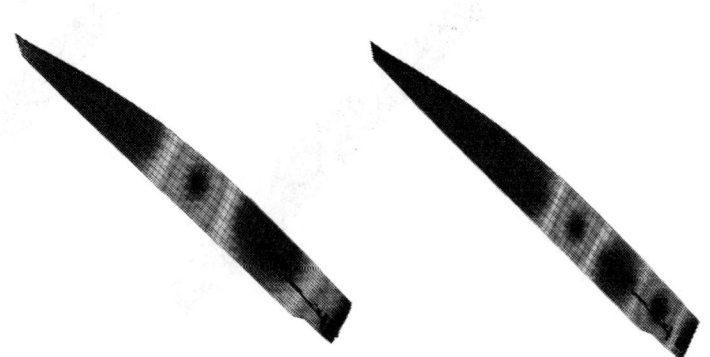

Table 23: Results obtained using the 1st, the 2nd, the 3rd and the 4th references cell as target values and two different sets of arrays.

Four reference cells

	7 Coating sensors	12 Coating sensors
Iterations	5	5
Objective function (mV^2)	530.41	242.42
Number of references in the constraints	0/4	0/4
SURFACE REF1: Potential (mV)	−856.0	−851.7
SURFACE REF2: Potential (mV)	−911.8	−907.0
SURFACE REF3: Potential (mV)	−839.4	−835.4
SURFACE REF4: Potential (mV)	−868.4	−870.5
Absolute error target, 1st Ref Cell (mV)	12.60	8.30
Absolute error target, 2nd Ref Cell (mV)	13.80	9.00
Absolute error target, 3rd Ref Cell (mV)	8.50	4.50
Absolute error target, 4th Ref Cell (mV)	10.50	8.40
Display		

Table 24: Results obtained using the 1st, the 2nd, the 3rd, the 4th and the 5th
references cell as target values and two different sets of arrays.

Five reference cells

	7 Coating sensors	12 Coating sensors
Iterations	5	5
Objective function (mV^2)	542.50	297.23
Number of references in the constraints	0/5	0/5
SURFACE REF1: Potential (mV)	−854.2	−851.2
SURFACE REF2: Potential (mV)	−911.1	−907.1
SURFACE REF3: Potential (mV)	−837.5	−835.9
SURFACE REF4: Potential (mV)	−866.8	−869.8
SURFACE REF5: Potential (mV)	−874.7	−875.7
Absolute error target, 1st Ref Cell (mV)	10.80	7.80
Absolute error target, 2nd Ref Cell (mV)	13.10	9.10
Absolute error target, 3rd Ref Cell (mV)	6.60	5.00
Absolute error target, 4th Ref Cell (mV)	12.10	9.10
Absolute error target, 5th Ref Cell (mV)	7.80	6.80
Display		

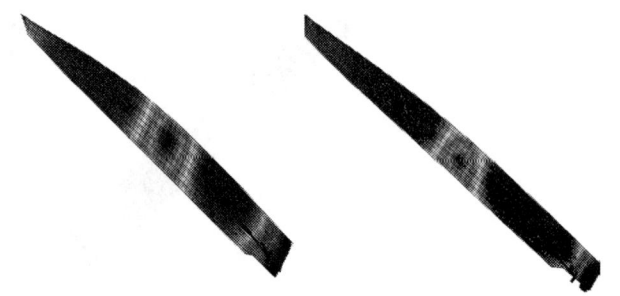

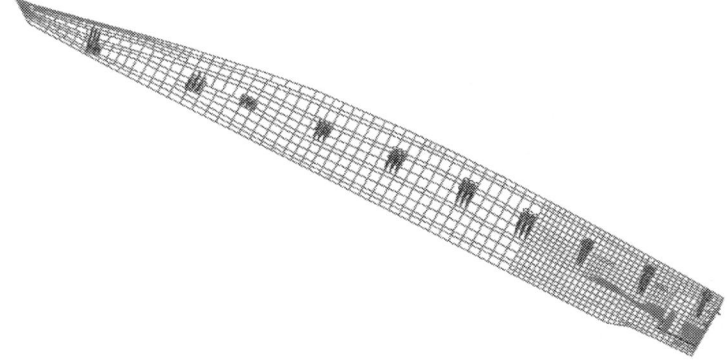

Figure 11: Ten coating sensors distribution on the surface of the frigate.

1.7 Summary

Two models have been analysed, a cylinder model and a frigate model. Both of them had three damaged areas on their surfaces. A compact radial basis function was employed since it becomes zero when the range is exceeded and, therefore, the damaged areas can be detected with respect to the coating area [8].

The objective function increases its value with the number of reference cells, since the distribution of the coating sensors and the interpolation does not exactly represent the real geometry of the damaged areas. This also results in the constraints not being satisfied when the number of reference cells increases. This would imply that the potential tolerance on the constraints should be increased as the number of reference cells increases. However, the solution has been found to converge (*i.e.* becomes accurate) as the number of reference cells increases in all the tests. There is a threshold above which the damage is effectively detected

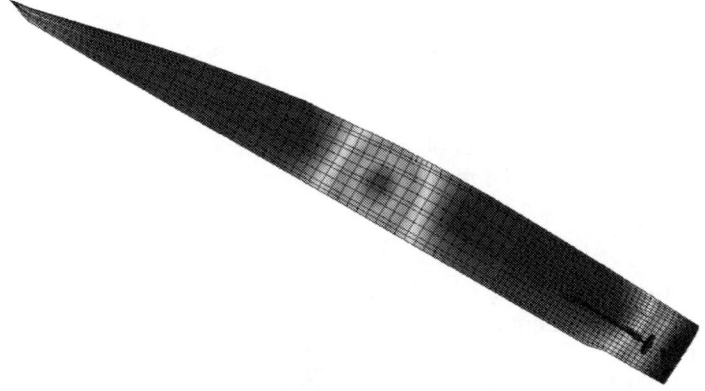

Figure 12: Prediction of the coating using five reference cells and 10 coating.

when enough reference cells data are provided, despite the constraints not being satisfied.

In general, the fact that the final optimum solution does not satisfy the constraints would be considered a failure of the method. However, experience with the techniques suggests that this is not the case and that the solution found is simply the best available. If the constraints are not satisfied and the targets are not matched, it does not mean that the solution is inaccurate but simply the closest one to the final target possible. Given the number and distribution of the sensors and interpolation method used.

The greater the number of coating sensors, the greater the capability of the optimisation to achieve the conditions imposed (constraints). Consequently, increasing the number of coating sensors makes the difference between the target potential and the final design potential becoming smaller.

1.8 Condition of the damaged areas

Despite the damaged areas found, there is no certain idea of what the condition of these damaged areas is. Knowledge of the current distributed to each one of the areas could indicate which of the areas is in the worse condition.

The prediction of the damaged areas is done by an interpolation method along the surface. Thus, there is no clear idea of the real shapes of these areas and therefore what amount of current is being taken.

Three sink points were placed on the top of each one of the damaged areas of the cylinder model, to try to predict this amount of current. Their currents and positions will be modified by the optimisation to achieve the same objective function (15) and constraints employed up to now. Four reference cells data were used in this experiment and the range of constraints are shown in the table 25.

None of the surfaces were in the constraints range by the end of the optimisation since the sinks employed could not emulate the real situation of the damaged areas.

The final positions of these three sink points are shown in fig. 13. The results indicate that the damaged areas predicted previously are confirmed by the sinks.

Table 25: Summary of results obtained when four reference cells were used as target value and 3 movable sinks to emulate the coating.

	Target	Constraints	% out target
Objective function	4134.82		
Number of references in the constraints	1/4		
SURFACE REF1: Potential	−829.2 −773.9	−770, −777	7.15
SURFACE REF2: Potential	−774.7 −770.3	−765, −775	0.57
SURFACE REF3: Potential	−784.0 −758.0	−755, −762	3.43
SURFACE REF4: Potential	−848.4 −867.8	−864, −870	−2.24

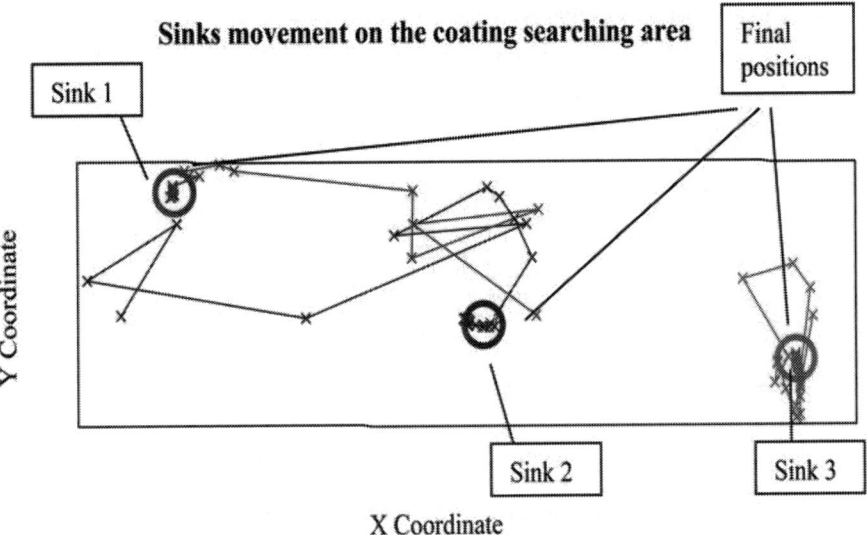

Figure 13: Final positions of the sink points along the search area.

The total current taken by the sinks after the optimisation is similar to the real total current taken by the damaged areas, with an error of 10.3%. However, the currents taken by each of the sinks do not follow the trend of the real current taken by the damaged areas (table 26). This is mainly due to the size and shape of the damaged areas and the difficulty to emulate them with the sink points.

These results could suggest the idea of using the information of the internal points underneath the cylinder to improve the prediction of the real distribution of current on the cylinder surface.

1.9 Condition of the damaged areas from UEP data

Data of the UEP (under electric potential) at 20m from the cylinder are provided. Three sinks have to be moved over the cylinder surface to confirm the positions previously predicted and predict the current taken by each one of the damaged

Table 26: Comparison between the real current taken by the damaged areas and the ones taken by the movable sinks when four reference cells were used.

	Current taken after optimisation (mA)	Real Current taken (mA)	% error
Sink 1	−11059.6	−5109.1	116.47
Sink 2	−2790.0	−8741.7	−68.08
Sink 3	−8756.8	−6650.1	31.68
Total	−22606.3	−20500.9	10.27

areas. The sum of least squares of the current densities with respect to the target values measured at the internal points are to be taken as objective function (18).

- Minimise the summation of the least squares of the components of the signature at the sensors beneath the vessel.

$$\text{Obj}_2 = \sum \left(I_{X_{t\,\arg et_i}} - I_{X_{computed_i}} \right)^2 + \sum \left(I_{Y_{t\,\arg et_i}} - I_{Y_{computed_i}} \right)^2$$
$$+ \sum \left(I_{Z_{t\,\arg et_i}} - I_{Z_{computed_i}} \right)^2 \tag{18}$$

Where:

$I_{X_{t\,\arg et_i}}$ is the X component of the target current density.

$I_{Y_{t\,\arg et_i}}$ is the Y component of the target current density.

$I_{Z_{t\,\arg et_i}}$ is the Z component of the target current density.

$I_{X_{computed_i}}$ is the X component of the computed current density.

$I_{Y_{computed_i}}$ is the Y component of the computed current density.

$I_{Zcomputed_i}$ is the Z component of the computed current density.

Two experiments were carried out: one considering the potential range at the reference cells, constraints; and a second one without considering the potential range at the reference cells (no constraints).

The optimisation will attempt to match the computed electric signature of the cylinder to the target one.

1.9.1 Without considering the reference cells

Three sink points were placed on the cylinder surface. Only the information related to the UEP is available, this is to say that just the current density in the three components of the electric field is available at a distance from the cylinder.

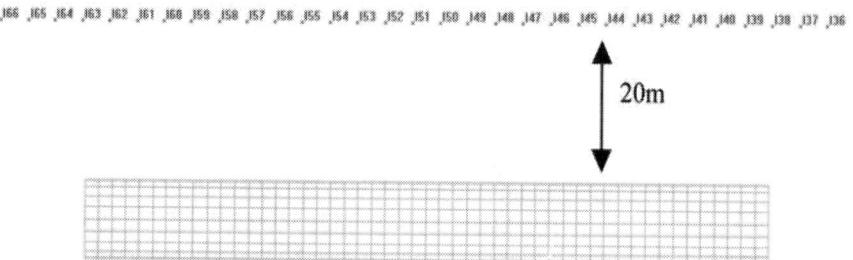

Figure 14: Internal points placed 20m from the cylinder.

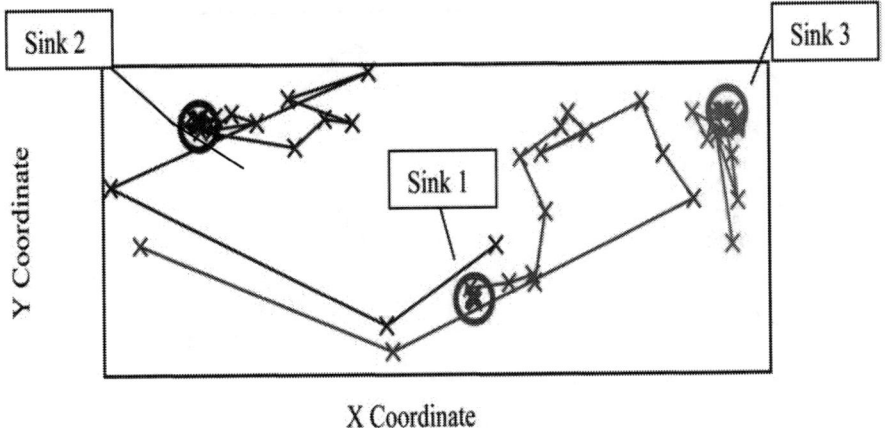

Figure 15: Final position of the sink points by using the UEP.

After the optimisation, the UEP computed has matched the target UEP. The final positions of these three sink points are shown in fig. 15. The results indicate that the damaged areas previously predicted are confirmed by the sinks.

The total current taken by the sinks after the optimisation differs by 29.2% with respect to the real total current taken by the damaged areas. In addition, the current taken by the sinks for the final design do not follow the trend of real current taken by the damaged areas as is shown in table 27. This confirms the hypothesis obtained in previous experiments, the sink points do not properly emulate the real size and shape of the damaged areas.

1.9.2 UEP and reference cells
In this case, not only the information related to the UEP is available but also the information related to the reference cells. A small potential range is set as constraints at the position where reference cells are located.

Table 27: Comparison between the real current taken by the damaged areas and the ones taken by the movable sinks when the UEP was used.

	Current taken after optimisation (mA)	Real Current taken (mA)	% error
Sink 1	−6577.1	−5109.1	28.73
Sink 2	−6301.19	−8741.7	−27.92
Sink 3	−13611.2	−6650.1	104.68
Total	−26489.5	−20500.9	29.21

Sinks movement on the coating searching area

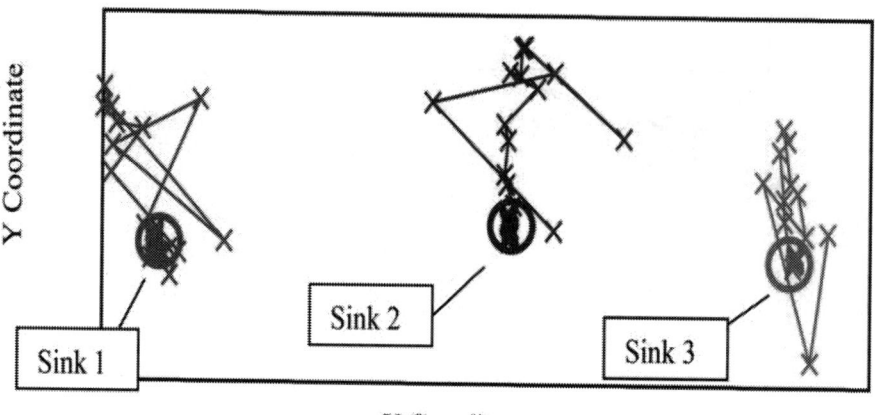

Figure 16: Final position of the sink points by using the UEP.

The optimisation will try to match the target UEP, keeping the potential of the reference cells within a small range around the measured value, smaller than 1%.

After the optimisation, the UEP has matched the target UEP.

The final positions of these three sink points are shown in fig. 16. As before, the results indicate that the damaged areas predicted previously are confirmed by the sinks.

The total current taken by the sinks after the optimisation is similar to the real total current taken by the damaged areas, with an error of 10.41% (table 28). On this occasion, the current taken by the sinks shows a major approach to the real taken current. The smallest damaged area, where the *sink 1* is placed, has been highlighted in this experiment. The results prove that with information about the UEP and potential measures at the reference cells an accurate prediction of the

Table 28: Comparison between the real current taken by the damaged areas and the ones taken by the movable sinks when the UEP and the reference cells potential were used.

	Current taken after optimisation (mA)	Real Current taken (mA)	% error
Sink 1	−5091.9	−5109.1	−0.34
Sink 2	−7884.9	−8741.7	−9.80
Sink 3	−9658.3	−6650.1	45.24
Total	−22635.1	−20500.9	10.41

current taken by the damaged areas could be achieved, despite the sink points not properly emulating the real size and shape of the damaged areas.

1.10 Three closest coating sensors, 4 reference cells

Two interpolation methods were considered to analyse the coating problem state, radial basis function and the three closest coating sensors. Up to now, the radial basis function has been the interpolation method studied in this work. The three closest coating sensors were also analysed in the frigate model with 5 reference cells (fig. 10).

An array of 26 coating sensors was placed on the frigate surface model (fig. 17). A triangular distribution of the coating sensors was considered to be the most suitable since the three closest sensors will be employed to interpolate the coating in the in-between spaces.

Figure 18 shows that damaged areas are reasonably predicted. However, this interpolation method was not considered as effective as the radial basis function since the solution depends on the correct distribution of the coating sensors. To obtain accurate results a fine triangular distribution of the coating sensors could become necessary.

1.11 Polarisation curve of the underlying material

Three polarisation curves were employed to represent different coating states of the surface. They were created by scaling the bare steel surface (fully uncoated) polarisation properties to obtain the 90% bare steel curve and the fully coated curve (fig. 19). Some tests were performed in order to test the difference

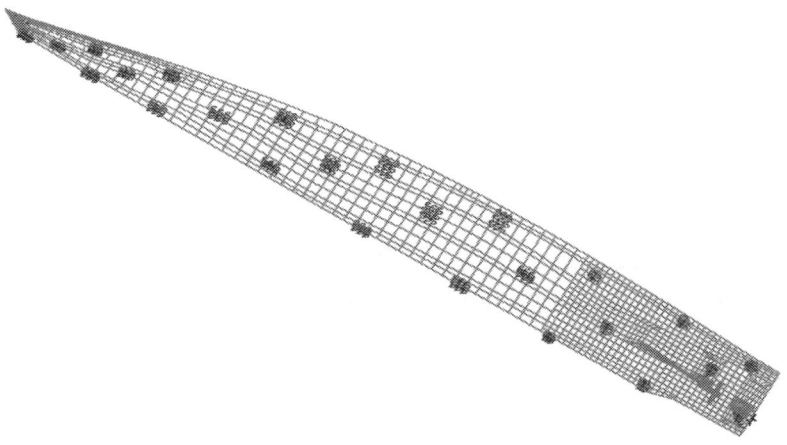

Figure 17: Coating sensors distribution on the prediction surface to study the three closest interpolation method.

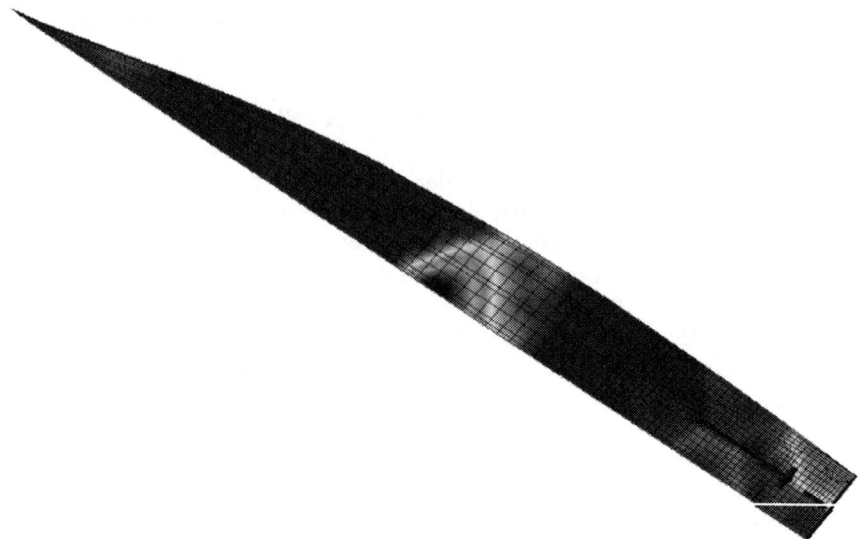

Figure 18: Prediction of the coating using five reference cells, 26 coating sensors and the three closest coating sensors as method of interpolation.

between using the real polarisation curve and a linear approximation. Figure 20 shows the real polarisation curve and a linear polarisation curve with only two

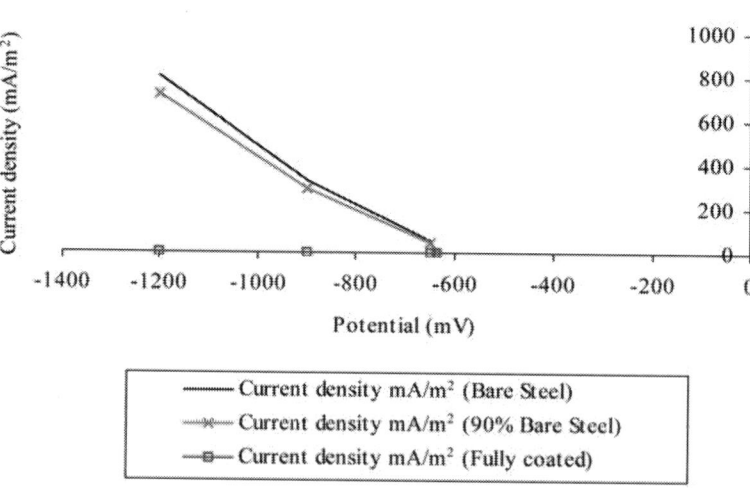

Figure 19: Polarisation curves for different grades of coating and underlying steel.

Comparison polarisation curves for different grades of coating,
with and without approximation

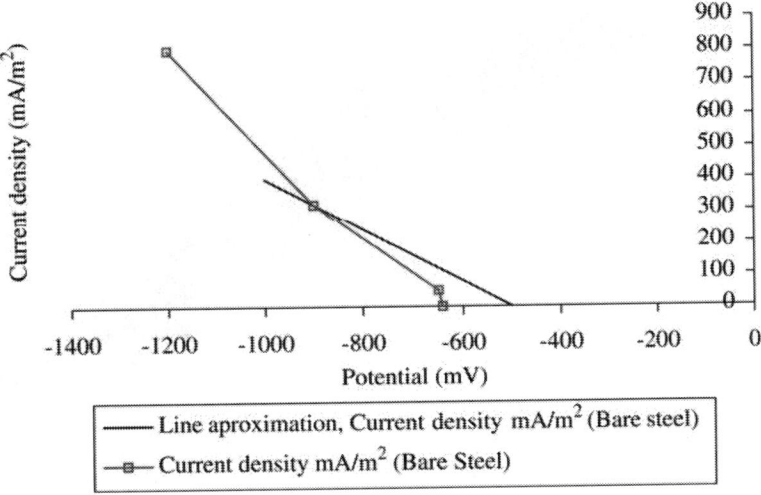

Figure 20: Comparison between the line approximation to bare steel and the real
polarisation data.

points, (–500mV, 0mA/m²) and (–1000mV, 400mA/m²) for bare steel in sea
water.

The coating state was predicted using a line which emulates bare steel and
another line with current near 0mA, which emulates the fully coated state. Five
reference cells were used (fig. 10) as target values and the set of twelve coating
sensors shown in fig. 9. The results obtained predicted the three damaged areas
in the approximate correct position (fig. 21) as when real polarisation data were
used (table 24).

1.12   Concluding remarks

The prediction of the position of the damaged areas on the surface of a frigate
model has been achieved by using the potential measurements at reference cells
on the structure.

In spite of being an approximation to the real state of the coating, the RBFs
or the three closest sensors enable the location of the damage to be predicted.

A minimum number of potential measurements are necessary to predict the
position of the damaged areas otherwise the prediction will not be accurate
enough and only some of the damaged areas will be revealed. However, an
increase in the number of coating sensors can improve the prediction from the
same target data.

Data from the corrosion related electric and magnetic fields can also be
employed to identify the condition of a vessel.

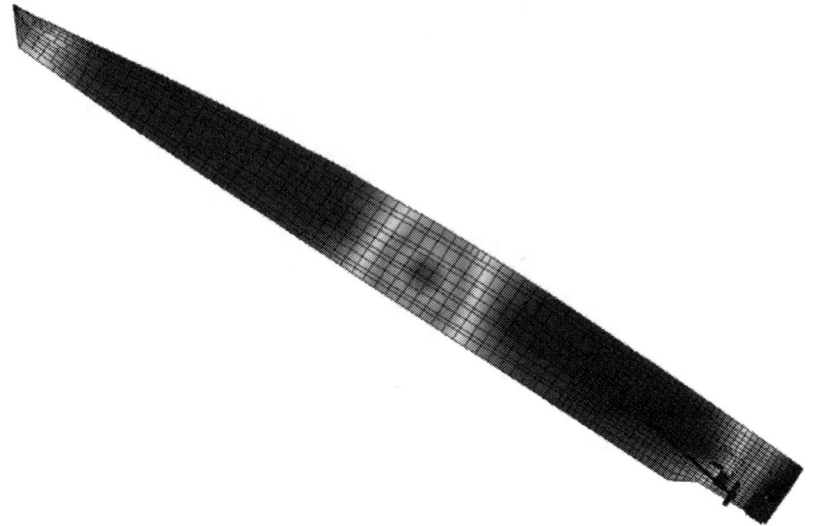

Figure 21: Prediction of the coating of the frigate by using five reference cells, 12 coating sensors and approximate polarisation data.

In the cases tested it was found that a linear approximation to the polarisation data (*i.e.* a curve which roughly represents the behaviour of the underlying materials) was sufficient to predict the location of the damage.

The methods presented could form the basis of a condition monitoring system or improved control system for CP systems.

In addition, the corrosion related electric and magnetic fields were also employed to identify the condition of the vessel.

The electric field and the potential measurements on the vessel can provide, with reasonable accuracy, the position and condition of the damaged areas.

Further testing is required on real shipboard data to validate the techniques further and draw up guidelines for the number of sensors and reference cells required.

References

[1]   Santana Diaz, E. & Adey, R., Optimisation of the performance of an ICCP system by changing current supplied and position of the anode, *Boundary Elements Methods* **24** Conference, Sintra, Portugal, 2002.

[2]   Santana Diaz, E., Adey, R., Baynham, J. & Pei, Y.H., Optimisation of ICCP systems to minimise electric signatures, Marelec conference 2001, Sweden, 2001.

[3]   Santana Diaz, E. & Adey, R., A Computational Environment for the Optimisation of CP system Performance and Signatures, WarshipCP Conference, Royal Military College of Science in Shrivenham, Cranfield University, United Kingdom, 2001.

[4]   Rawlins, R.G., Davidson, S.J. & Wilkinson, P.B., Aspects of corrosion related magnetic (CRM) signature management, Wembley Conference Centre, London, UK, 1998, 237–241.

[5]   Rawlins, R.G., Davidson, S.J. & Wilkinson, P.B., Aspects of corrosion related magnetic (CRM) signature management, Wembley Conference Centre, London, UK, pp. 237–241 (1998).

[6]   Aoki, S., Amaya, K. & Gouka, K., Optimal cathodic protection of ship, In *Boundary Element Technology XI*, ed. R.C. Ertekin, C.A. Brebbia, M. Tanaka & R. Shaw. pp. 345–356 (1996).

[7]   Diaz, E.S. & Adey, R., Corrosion optimisation using boundary elements, *Boundary Elements Communications;* **12**, N.1, 2001; 12–25.

[8]   Leitao, V.M.A., & Tiago, C.M., The use of radial basis functions for one-dimensional structural analysis problems, *Boundary Elements XXIV*, eds. C.A. Brebbia, A. Taden & V. Popov. pp. 165–179 (1996).

[9]   Vanderplaats, G.N., DOT/DOC Users Manual, Vanderplaats, Miura and Associates, 1993.

[10] Castillo, E., Conejo, A.J., Pedregal, P., García, R., & Alguacil, N., *Building and solving mathematical programming models in engineering and science*, Wiley Interscience, 2001; 190–207.

[11] Morgan, John, *Cathodic Protection*. National Association of Corrosion Engineers, NACE. (1987).

Numerical simulation of the cathodic protection of pipeline networks under various stray current interferences

L. Bortels & J. Deconinck
Department of Electrical Engineering (TW-ETEC), Vrije Universiteit Brussel, Belgium.

Abstract

A simulation software for the design and evaluation of the cathodic protection (CP) of underground pipeline networks is presented. The macroscopic model behind the software expresses Laplace's equation in a flat and homogeneous three-dimensional half-space (soil) and takes into account the non-negligible ohmic voltage drop along the pipes. The soil (external problem) and pipes (internal problem) are coupled with each other through the pipeline coating. For this coating, an advanced model has been developed that takes into account the local soil resistivity, the holiday ratio, the average holiday size, the coating thickness and the coating resistance.

The model is solved using a combination of the boundary element method (BEM) for the external problem and the finite element method (FEM) for the internal problem. Special 'pipe-elements' have been used to reduce the calculation time. Anodes and railway tracks are replaced with equivalent (half-)pipes having the same resistance-to-soil as the anode or track, allowing a straightforward and flexible integration of all components (pipes, anodes and tracks) in the same model. Finally, rectifiers, current drains, sub-stations, electric trains and others are modelled as equivalent electrical networks that can connect any two points of the total configuration.

As a result, the model gives the 'on' and 'off' potential along the pipe, the potential distribution in the soil, the axial currents flowing through the pipes as well as the radial current densities leaving or entering the pipe walls. The software can deal with all standard kinds of cathodic protection interferences as well as stray currents coming from other CP-installations, DC-traction systems, HVDC-transmission lines, grounding systems and others.

1 Introduction

Failures in oil or gas pipelines can have severe environmental and economic consequences. Therefore, large investments have been made in studies on corrosion prevention for buried pipes. Important research is being conducted to determine and predict the corrosiveness of the soil, corrosion mechanisms in the ground and effective protection techniques such as coating and painting of menaced metallic structures. Moreover, because of the hidden character of pipelines and their low accessibility, installation, survey, maintenance and repair is intricate, elaborate and expensive.

Numerical modelling can provide some relief by simplifying and optimizing installation, maintenance and repair. When used in the planning phase, conceptual mistakes can already be traced before any actual installation, by calculating different set-ups in cheap, harmless and fast simulations. Also, a model can provide reference values for measurements on operational sites, that can help in the tracing and solving of any possible anomaly. Last but not least, the model technique creates a safe and cost effective on screen 'virtual' test environment where new corrosion engineers can gain experience without long and expensive 'trial and error' experiments on site.

The fundamentals of such a mathematical model have been developed by F. Brichau, J. Deconinck and T. Driesens [1, 2]. Recently, ElSyCa developed CatPro [3], a PC-version of the code with windowing user-interface. (ElSyCa is a spin-off company of the Vrije Universiteit Brussel, developing simulation software in the field of electrochemistry and cathodic protection.)

The basic ideas and all fundamental aspects of the model are explained here. Mathematical details and the validation of the basic model have been discussed in detail in papers [1, 2].

2 Design of CP-systems : 'on' and 'off' pipe-to-soil potentials

When designing a cathodic protection system, the aim is to obtain a pipe-to-soil potential along the developed length of the pipeline network that is more negative than a certain minimum protection level. In soils, this minimum level is normally taken at -0.85V versus a copper-sulphate reference electrode (CSE) that needs to be placed directly adjacent to the pipeline in order to reduce the IR-drop in the soil and over the coating. The value obtained in this situation is referred to as the 'off' potential.

In practice however, due to the hidden character of the pipeline, it is often not possible to put the reference electrode directly near the pipeline. Instead, the reference electrode is put at the soil surface, directly above the pipeline which can result in important IR-drop errors. The value obtained here is referred to as the 'on' potential. In normal operating conditions, this value is more negative than the (true) 'off' potential, resulting in an overestimate of the obtained protection level.

In the following section, the mathematical model will be introduced. It will be demonstrated that both the 'on' and 'off' potentials can be calculated giving valuable information on the obtained protection levels of the CP-configuration, even in very complex interference situations.

3 The mathematical model coupling BEM and FEM

In order to describe properly vast buried pipeline structures, one of the basic ideas of the model is to link the 'external' world — the soil — with the 'internal' world being the metallic conductor of the pipe (see fig. 1).

3.1 Soil and pipes — the external domain

The soil together with the pipes and anodes, being also a kind of pipes, is to be considered as an electrical/electrochemical system in which the earth acts as an electrolytic solution (conducting medium) and the outer metallic pipe surfaces are electrodes. If no ion concentration gradients are considered in the soil, Ohm's law applies and the current density J follows from:

$$\vec{J} = -\sigma\vec{\nabla}U, \tag{1}$$

where U is the potential, σ is the electrical conductivity of the soil and $\vec{\nabla}U$ is the electric field. When the conductivity of the soil is supposed to be uniform, the potential distribution in the soil is described by the Laplace equation, expressing conservation of charge:

$$\vec{\nabla}\cdot\left(\vec{\nabla}U\right) = 0. \tag{2}$$

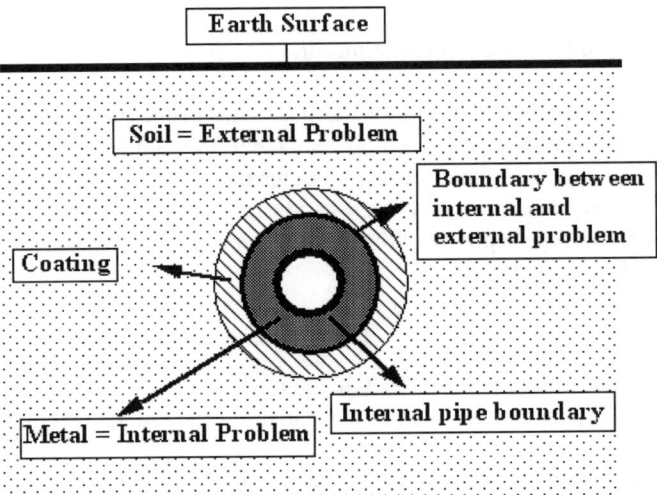

Figure 1: Cross section of a buried pipeline.

Do remark that at the earth surface (insulating boundary) the current density normal to the boundary is zero.

Because of the complex geometries found in practice and the non-linear boundary conditions, the Finite Element Method (FEM) and the Boundary Element Method (BEM) [4] have been the most successful numerical techniques to solve Laplace problems. However, there are strong arguments for using BEM to model underground corrosion. First of all, BEM is intrinsically well suited for infinite problems, since only the inner boundaries are to be defined with the infinite border implemented inherently. Further, assuming a perfectly flat earth surface, based on the so-called 'mirror-point' technique, the earth surface needs not to be discretised. Last but not least, the Boundary Element Method provides directly the local current densities on the metal surfaces, responsible for the corrosion.

3.1.1 Introduction of pipe elements

Because of the specific cylindrical shape of pipelines, special 'pipe elements' are used to perform the discretisation. These elements suppose a uniform, radial current density distribution. From a numerical point of view, this considerably reduces the problem description, since data input is based on outer pipe diameters and pipe lengths while elaborate element meshing is avoided. When one studies the error involved by assuming a uniform radial current density, it appears that the use of these pipe elements is justified, except when the configuration is bare and narrow, when coating holidays appear or when it is very close to another structure. Consequently, these hypotheses are not a true restriction in most cases encountered in reality.

In the same way, anodes will be replaced with a cylindrical rod having the same resistance-to-earth as the original groundbed as will be outlined in section (5.1). In fact formula, giving the resistance of a specific groundbed, usually are derived by making very similar assumptions.

3.2 Metallic pipes — the internal domain

For vast underground pipe structures the resistivity of the pipes in axial direction is no longer negligible. For a given pipe segment, the radial current coupled with the external world, enters and leaves the outer pipe surface (through the coating) while the collected axial current changes and causes a potential drop. The potential drop ΔV across a pipe segment with length L can be described using the following mathematical expression:

$$\Delta V = R_p \left\{ I_{ax} + \int_0^L J dS \right\}, \tag{3}$$

in which V is the potential of the metallic part of the pipe, R_p the resistance of the pipe segment, I_{ax} the collected axial current and J the radial current density.

3.3 Coupling between external and internal world

The coupling between the external problem (U) and the internal problem (V) is simply achieved by expressing the boundary condition on the external surface of each pipe segment giving:

$$V - U = \eta(J_b) + RJ + E_{corr},\qquad(4)$$

in which U is the potential in the soil adjacent to the pipe, $\eta(J_b)$ is the pipe polarisation, RJ is the voltage drop across the applied coating (if any) and E_{corr} is the corrosion potential of the metal-soil system. The relation between the 'macroscopic' current density J and the current density J_b through the bare steel is given by:

$$J = \theta J_b,\qquad(5)$$

in which θ represents the fraction of bare steel.

3.3.1 Polarisation of bare steel

The relation between the pipe polarisation and bare steel current density as used here is based on the work done by Orazem *et al* [5, 6]. They developed a fundamental expression for the electrochemical reactions that can occur on bare steel. An overview of the reactions that have been included in this work, together with the derived mathematical expression for the (partial) current density of that reaction, is given below:

- oxidation of metal : $Fe \rightarrow Fe^{2+} + 2e^-$

$$J_1 = 10^{(\eta^* - E_{Fe})/\beta_{Fe}},\qquad(6)$$

- reduction of oxygen : $O_2 + 2H_2O + 4e^- \rightarrow 4OH^-$

$$J_2 = -\frac{1}{\dfrac{1}{-J_{lim,O_2}} + 10^{(\eta^* - E_{O_2})/\beta_{O_2}}},\qquad(7)$$

- hydrogen evolution : $2H_2O + 2e^- \rightarrow H_2 + 2OH^-$

$$J_3 = -10^{-(\eta^* - E_{H_2})/\beta_{H_2}},\qquad(8)$$

with all parameters as explained below:
- $\eta^* = V - U^*$, the polarisation of bare steel,
- V the potential of the metallic part of the pipe,
- U^* the potential near the pipe just under the coating,
- J_1, the partial current density due to metal oxidation,
- E_{Fe}, the 'effective' equilibrium potential of Fe-reaction including the effect of the exchange current density,
- β_{Fe}, the Tafel slope for the Fe-reaction,
- J_2, the partial current density due to oxygen reduction,
- E_{O_2}, the 'effective' equilibrium potential of O_2-reaction including the effect of the exchange current density,

- β_{O_2}, the Tafel slope for the O_2-reaction,
- J_{lim,O_2}, the mass transfer limited current density for the O_2-reaction,
- J_3, the partial current density due to hydrogen evolution,
- E_{H_2}, the 'effective' equilibrium potential of H_2-reaction including the effect of the exchange current density,
- β_{H_2}, the Tafel slope for the H_2-reaction,

The total current density is given by:

$$J_b = J_1 + J_2 + J_3 = g(\eta^*), \qquad (9)$$

a strong non-linear function of η^*. Remark that formula (9) is not relative to the corrosion potential E_{corr}. The corrosion potential is that value for η^* that is obtained when the current density, calculated using equation (9), is zero.

3.3.2 Coating quality

The potential drop across the coating is given by the relation:

$$U^* - U = RJ, \qquad (10)$$

with:
- U the potential in the soil adjacent to the pipe,
- R the coating resistance.

The model for the coating as used here describes a perfect coating (no holidays) with a number of distributed holidays. All holidays are supposed to be cylindrical, with the same height as the coating thickness and with the same (average) diameter. In addition, it is assumed that these cylindrical holidays are filled with soil. According to literature [7], the resistivity of the holiday in first approximation is equal to about 10% of the surrounding soil resistivity.

The parameters for the coating resistance as described above are listed below:
- coating thickness,
- coating resistance,
- holiday fraction,
- resistance of holiday (as a fraction of local soil resistivity),
- average holiday diameter.

Based on these parameters and the local soil resistivity, the total coating resistance R can be calculated. Do remark that in the current model, the soil resistivity can be specified for each individual segment of the pipeline network.

3.3.3 Outcome of the calculations

Due to the coupling of internal and external domains the boundary conditions are only to be applied on the metallic structures (anodes and pipes). Either the total currents entering and leaving the whole system at given points, or the potential difference(s) between two or more points of the structure, or combinations of both are imposed. The general approach allows to deal in a straight forward way with two or more non intentionally connected metallic structures. They are anyway coupled via the earth.

As a consequence, the software provides all data of interest from a corrosion point of view:

- the 'off' potential in all points of the pipeline, by calculating the voltage difference between the (steel) pipe and the underside of the coating just above the steel,
- the 'on' potential in all points of the pipeline, by calculating the voltage difference between the (steel) pipe and the soil at surface level just above the pipeline,
- the axial currents flowing through the pipeline walls,
- the radial current densities entering or leaving the pipeline surfaces.

In addition, the ground resistivity, metal resistivity and geometrical aspects of the pipelines and anode beds, polarisation characteristics of the metallic surfaces, coating resistivity as well as imposed voltages and/or currents, joints, insulating joints and current drains are considered.

4 Modelling stray currents

In practice, stray currents play a non-negligible role in many underground cathodic protection problems. With respect to a given structure, a stray current is to be defined as a current flowing on a structure that is not part of the intended electrical circuit. For corrosion to occur as the result of stray currents, there must be an exchange of current between a metallic structure and an electrolytic environment. Due to stray currents, interference occurs.

Sources of stray currents are manifold. Some are caused by other cathodic protection installations, grounding systems, welding posts (and others), referred to as steady state stray currents. But most often traction systems like railroads and tramlines are responsible for large dynamic stray currents. As a result, also the prevention techniques for pipelines are numerous going from painting and coating to traditional cathodic protection and current drainage.

Similar to conventional corrosion prevention, all these methods reduce the stray current without increasing the anodic current density. Otherwise, high local anodic currents would cause rapid failure. In what follows these aspects will be considered in more detail.

4.1 Steady state stray currents

Because of the general fundamentals of the model, stray current problems involving different cathodic protection systems — with current or voltage controlled groundbeds — can be simulated without any preliminary extension. Even stray currents arising from grounded systems, like welding installations, may be calculated. The influence of distinct cathodic protection systems on each other appear automatically.

The well-known phenomenon of protection currents that bring about stray current corrosion in neighbouring bodies is called 'interference'. According to the kind and origin of the potential variations that come into play (cathodic or anodic), a distinction is made between cathodic, anodic, combined and induced interference. They all arise in a natural way in the model.

4.2 Dynamic stray currents

The initial approach of the underground model was exclusively based on buried cylindrical structures. In order to quantify stray-currents coming from traction systems, these traction systems are also to be modelled. This means that the rails and the above ground part containing sub-stations, overhead wires and trains are to be considered.

For reasons of logics and efficiency, the overhead network is considered as a separated electrical network, consisting of conductors, characterised by finite axial resistivity (lifelines, etc.) and node potentials (at sub-stations, trains, etc.) but with infinite radial resistivity since no current can leave the wire through the air. In this way, from the mathematical point of view, the overhead network is treated in just the same way as the internal pipe problem.

On the other hand, the rails have an axial and a radial resistivity, such that current can flow to and from the soil. A traction current that leaves the rails becomes a stray current that can enter a protected system. This means that, from the electrical point of view, there is nearly no difference between the buried pipelines and the rails at the earth surface.

4.3 Introduction of 'rail-pipes'

Although the shape of the rails together with bed and sleepers is not cylindrical, an equivalent semi-cylindrical 'pipe' can be defined that is a fairly good approximation, certainly at some distance from it. The axial resistivity is that of the rails and the radial resistivity corresponds to the resistivity found between rails and ground (details are found in section 5.2). So, in accordance with the standard definition of pipe elements in the model, the tracks are represented as equivalent pipes ('rail-pipes'). Their axis coincides with the earth surface and they are only in contact with the ground with their lower half. The axial 'rail-pipe' current equals that part of the traction current that directly returns to the sub-station. As the rail-pipes are coupled with the earth, the radial current densities can become stray currents when they are picked up by structures.

The 'rail-pipes' are coupled with the overhead network through the sub-stations. A given voltage difference is imposed between both. Finally, trains act as loads between overhead wires and 'rail-pipes'. They can be put on any place and the different traction modes can be simulated by varying the voltage (emf) and resistance of the load. All this has been indicated in figs 2 and 3.

This uniform approach makes the data entry very transparent and straightforward. The identification of the different types of involved rails, overhead wires, sub-stations, buried pipelines and protection systems can be based on their specific spatial and/or electrical characteristics.

5 Converting anodes and tracks into equivalent pipe elements

As mentioned before in sections (3.1.1) and (4.3), special 'pipe elements' are used to describe all pipes, anodes and tracks in the CP-configuration. Pipe elements have

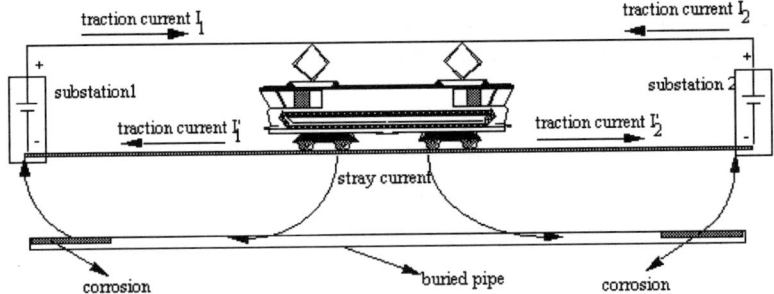

Figure 2: Traditional stray current situation for a railway.

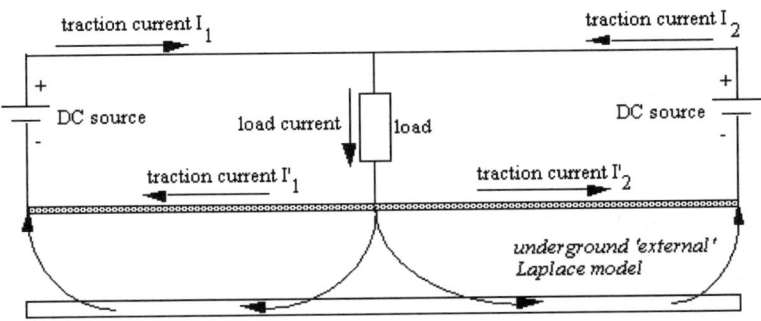

Figure 3: Equivalent electrical network used in the software.

uniform radial properties (current density, potential, ...) and are described by their outer diameter, inner diameter (or wall thickness) and axial resistivity.

For pipes, due to their intrinsic cylindrical shape, these data are directly available and no transformation formulas are needed. Anodes and tracks on the other hand can have shapes that are far from the *ideal* cylindrical shape. Therefore, special transformation formula are applied to obtain the equivalent 'pipe element'.

5.1 Converting an anode into an equivalent pipe element

Except when near anode details are needed, any anode bed can be replaced by an equivalent vertically buried full rod that has the same earth resistance as the original anode bed. For a full rod ($D_{out} = D, D_{in} = 0$), this resistance R is given by the well-known formula:

$$R = \frac{\rho_{soil}}{2.\pi.L} \ln \left[\frac{4L}{D} \right],$$ (11)

with L the length of the rod, D the diameter of the rod and ρ_{soil} the soil resistivity. The use of this formula is justified if $L \gg D$ and if there are no other electrodes in a region $< 5L$.

When doing numerical calculations, the anode with length L is divided into a number of calculation elements. The minimum length L_{min} of such an element is bounded by the condition:

$$L_{min} > 3D, \tag{12}$$

as outlined in the work of Modjtahedi *et al* [8]. Indeed, too small elements can cause oscillations in the calculated potential distribution along the anode. The non-linear equation (11) can be simplified by introducing a direct relation between L and D as given below:

$$L = kD, \tag{13}$$

with $k \gg 1$ in order to obey relation (12).

Combining equations (11) and (13) finally gives the length L of the equivalent anode element as a function of the resistance:

$$L = \frac{\rho_{soil} \cdot \ln(4k)}{2.\pi} \cdot \frac{1}{R}. \tag{14}$$

To conclude, the equivalent resistance R of the anode bed is obtained by dividing the measured anode potential (measured with respect to the far field) V_a and the rectifier current I:

$$R = \frac{V_a}{I}. \tag{15}$$

5.2 Converting a track into an equivalent pipe element

As outlined before in section (4.2), a track has to be replaced by an equivalent semi-cylindrical 'pipe element'. The original couple of rails that form the track have cross-section S. Common practice is to set the outer diameter D_{out} of the equivalent half-pipe equal to the distance d_{rails} between the center of the axes of both rails:

$$D_{out} = d_{rails}. \tag{16}$$

The inner diameter D_{in} is found by expressing that the cross-section of the equivalent half-pipe equals that of both rails:

$$\frac{\pi[D_{out}^2 - D_{in}^2]}{8} = 2S, \tag{17}$$

or in other words:

$$D_{in} = \sqrt{D_{out}^2 - 16\frac{S}{\pi}} \tag{18}$$

The axial resistivity ρ of the equivalent half-pipe is of course equal to that of the rails (assuming that both rails are made out of the same material).

6 Example 1: a simple pipe-anode configuration

The first example investigates the cathodic protection of a simple pipe-anode configuration.

6.1 Overview of the problem

Consider the problem of fig. 4. A coated steel pipeline buried at 5m depth receives cathodic protection from a long line zinc anode lying at the same depth by means of an impressed current system ($CS1$). The soil resistivity is 100Ω.m.

The pipe has an outer diameter of 50cm, a wall thickness of 1cm and a coating with an average resistance of 220Ω.m^2 and holiday ratio of 1%. The polarisation curve for bare steel has been taken from literature [6] and is presented in fig. 5 (with respect to the corrosion potential being -0.56V versus CSE). The anode is a solid cylindrical rod with a diameter of 10cm and an equivalent 'polarisation resistance' of 1Ω.m^2. The corrosion potential is -1V. Both the pipe and anode have a length of 100m.

6.2 Calculated 'on' and 'off' potentials

The calculated 'on' and 'off' potential are presented in fig. 6. As expected, the 'on' potential predicts higher protection levels than actually achieved along the pipeline. This is due to the additional effect of the IR-drop in the soil and across the coating which has not been eliminated (do remark that for the 'on' potential calculation the reference electrode is placed at the soil surface with all rectifiers on).

The error made when considering the 'on' potential as the protection criterion instead of the 'off' potential is about 177mV! The voltage drop across the coating

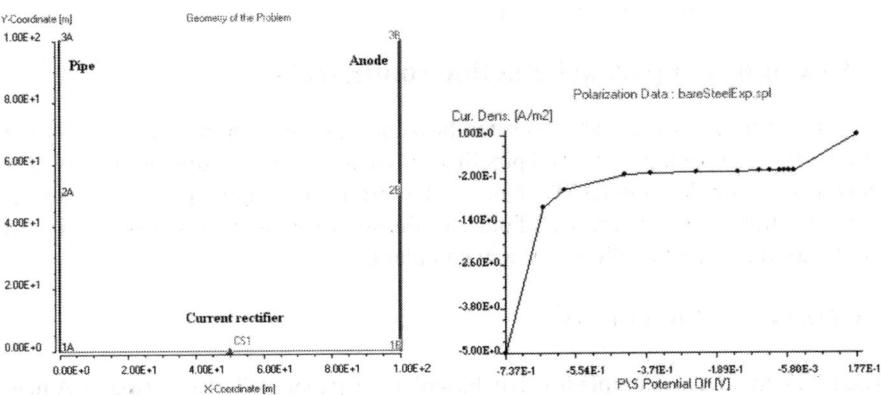

Figure 4: Geometry of the pipe-anode configuration (top view).

Figure 5: Polarisation curve of bare steel (relative to E_{corr}).

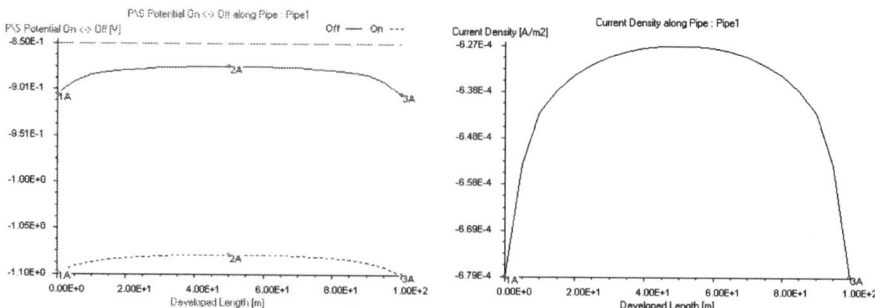

Figure 6: 'On' and 'off' potential along   Figure 7: Radial current density along
        the pipe.   the pipe.

(current density multiplied with coating resistance) is about 147mV while that in the soil is about 30mV. A plot of the current density is presented in fig. 7. It must be mentioned that, since the fraction of bare steel is 1%, the 'macroscopic' current density J presented here is 100 times smaller than the actual polarisation current.

6.3 Potential distribution in the soil

The soil potential distribution in the neighbourhood of the pipe-anode system has been calculated on a raster with corners $(-50.0, -50.0, 0.0)$ and $(150.0, 150.0, 0.0)$, and 31 points in both directions. The potential isolines calculated at earth level are presented in fig. 8. A maximum potential level of 69.2mV is obtained in the centre above the anode.

Similar results can be found by calculating the soil potential along a cutting line. This line goes from point $(-50.0, 50.0, 0.0)$ to point $(150.0, 50.0, 0.0)$, resulting in the soil potential distribution at earth surface level above the centre of the pipe-anode system. Results are shown in fig. 9. The same minimum and maximum values for the soil potential as before are found.

7 Example 2: a parallel pipeline configuration

In this example it will be demonstrated how the software can be used to study the interference between two parallel pipelines having different coating properties. It is shown how a single groundbed can be used to protect multiple pipelines. Finally, it is demonstrated how experimental data for the soil resistivity along the pipeline(s) can be used as input for the numerical calculations.

7.1 Overview of the problem

The idea behind this example is partly based on the paper of Carlson *et al* [9]. A new 36″ pipeline (referred to as 'Main Pipe' or 'New Pipe') with a total length of 84km is coated with fusion-bonded epoxy (FBE). Along its total length, it runs parallel to an older 24″ pipeline ('Secondary Pipe' or 'Old Pipe') at a constant distance of 20m.

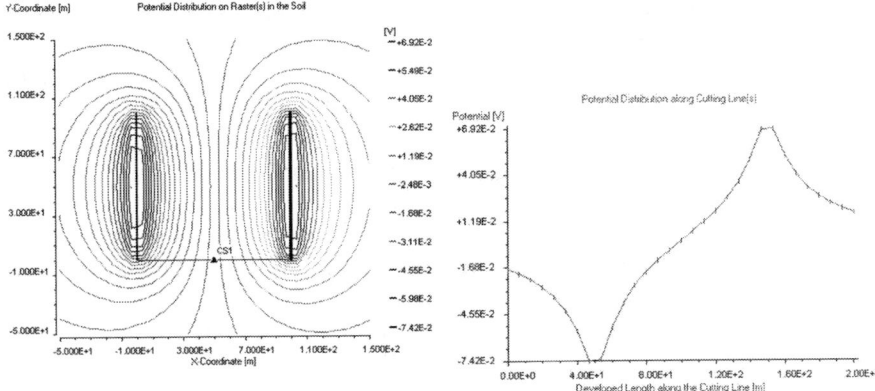

Figure 8: Soil potential distribution in the vicinity of the pipe-anode system.

Figure 9: Soil potential distribution above the centre of the pipe-anode system.

The centre of both pipes is located at a constant depth of 1.5m. The average soil resistivity along the pipeline trajectory is 20Ω.m.

The polarisation behaviour for both pipes is described by a non-linear polarisation curve that takes into account metal oxidation and oxygen and water reduction. The corrosion potential for the new and old pipe is -0.656V and -0.704V, respectively (versus CSE).

The model for the coating quality used for both pipes describes a perfect coating (no holidays) with a number of distributed holidays. The holiday ratio is 0.005% for the new pipeline and 0.1% for the old one. Taking into account a soil resistivity of 20Ω.m, this yields a coating resistance of about 31300Ω.m^2 and 1570Ω.m^2, respectively.

7.2 Cathodic protection design analysis

Based on the soil resistivity and coating quality data as specified in the previous section, an average current density distribution of $1.25 \ 10^{-5}$A/m^2 for the new pipeline is taken. This results in a current requirement of 35.9mA/km or about 3A in total.

For the old pipe, having a coating resistance that is about 20 times lower than that of the main pipe, an average design current density of $2.2 \ 10^{-4}$A/m^2 is taken. Hence, the old pipe requires a current value of 420.2mA/km or about 35A in total.

As outlined in the work of Carlson *et al* [9], and taking into account the assumed coating quality as described above, the groundbed spacing for the new pipeline is about 30km. With the total length of the pipe being 84km, the minimal number of groundbeds to ensure a smooth voltage profile is 3, resulting in a final spacing of 28km. In terms of developed length, this means that groundbeds should be placed at 14, 42 and 70km. Since the total current requirement for the main pipeline is 3A, a current output of 1A per groundbed is needed.

For the old pipeline, having an inferior coating and hence an increased attenuation, an extra groundbed in-between two successive groundbeds is placed (at 7, 28, 56 and 77km). This means that for the old pipe in total 7 groundbeds are used, each delivering a current output of 5A. Do remark that for the old pipe, to be fully 'correct', the 7 groundbeds should be placed with a spacing of 12km and not 14km as it is now (resulting in a smaller spacing and hence a higher protection level at begin and end).

In this example, we will use deep-anode groundbeds, located at a depth of 70m and at a horizontal distance of 10m with respect to the pipe. The resistance-to-ground of the anode beds is 1Ω.

An overview of the pipeline geometry and rectifier location is given in fig. 10.

7.3 Results

First of all, one will verify the design of each pipeline separately. Finally, the complete CP-configuration and the interference between both pipes will be investigated.

7.3.1 Study of the CP of the main pipeline

What we are interested in, is the 'off' potential along the developed length of the pipeline. As can be seen from fig. 11, the design as outlined above gives a very smooth distribution with pipe-to-soil values ranging between −0.90 and −0.91V. Figure 12 gives a comparison between the 'on' and 'off' potentials. The IR-drop error made when considering the 'on' potential is around 410mV.

From the solver output presented in fig. 13 and the current density profile from fig. 14, it can be noted that for a current output of 1A per anode, the average current density along the pipeline matches the designed value (see 7.2).

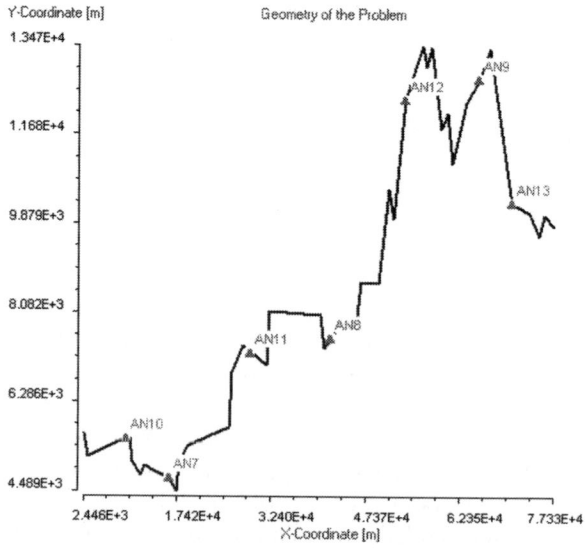

Figure 10: Position of the rectifiers along the pipelines (top view).

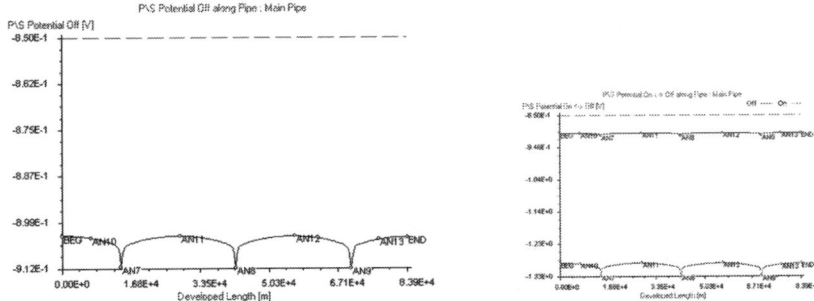

Figure 11: 'Off' potential for the main pipeline.

Figure 12: Comparison between 'on' and 'off' potential (main pipeline).

7.3.2 Study of the CP of the secondary pipeline

As outlined before, the secondary pipeline is protected using 7 groundbeds, each delivering a current output of 5A and with the labels for the rectifiers being AN7 to AN13.

As can be seen from fig. 15, the design as outlined above again gives a smooth distribution with 'off' values ranging between -0.92 and -0.95V. The calculated 'on' potentials are again much more negative than the corresponding 'off' potentials (fig. 16). The highest IR-drop (555mV) is found near groundbed AN13, the lowest one (340mV) in the middle of AN8 and AN12.

However, when compared with the 'off' potentials of the main pipeline, it can be noticed that at the beginning and end of the pipeline, there is a slight 'overprotection' when compared with the rest of the pipeline. This effect is due to the groundbed location for the secondary pipeline which is not 'perfect' as already mentioned in section 7.2.

7.3.3 Study of the CP of both parallel pipelines

Do remark that for the parallel pipeline problem, groundbeds AN7 to AN9 will be used to protect both pipelines as can be seen from fig. 17. One side of the

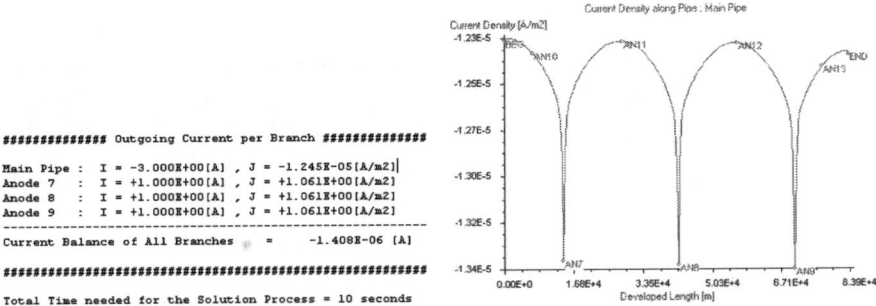

Figure 13: Solver output (main pipeline).

Figure 14: Current density (main pipeline).

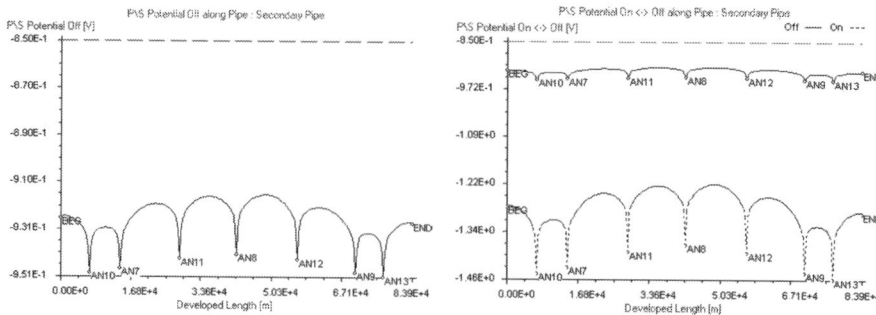

Figure 15: 'Off' potential for the secondary pipeline.

Figure 16: 'On' and 'off' potential for the secondary pipeline.

current rectifier is connected to the anode while the other side (split point − SP1) is connected to both pipelines using two different connections (JO1 and JO2). The total rectifier current (being 1A + 5A = 6A) will be distributed between both pipelines depending on the coating resistance, polarisation and ohmic effects in pipes and soil.

As can be seen from the solver window (fig. 18), the total protection current for the main pipeline is 3.1A, 34.9A for the secondary pipeline and 38.0A in total. This means that, due to the rectifier connection of fig. 19 an additional current of 0.08A is used for the cathodic protection of the main pipe. Indeed, the currents flowing in the external connections of the main pipeline (fig. 19) at locations AN7 to AN9 are respectively equal to 1.3, 0.4 and 1.4A. The other part of the total current of 6.0A at those locations is consumed by the secondary pipeline as seen from fig. 20.

The 'off' potentials along both pipelines are presented in figs 21 and 22 and compared with those calculated from the separate configurations. As before, the distribution is rather smooth with peak values of −0.90 and −0.94V for the main pipeline and −0.92 and −0.95V for the secondary pipeline.

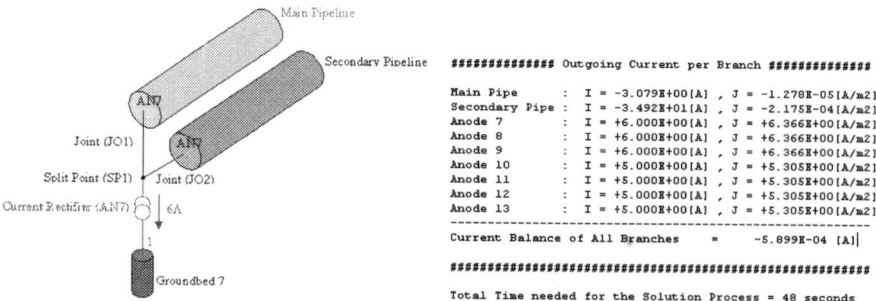

Figure 17: Protecting two pipelines using a single groundbed.

Figure 18: Solver output for the parallel pipeline configuration.

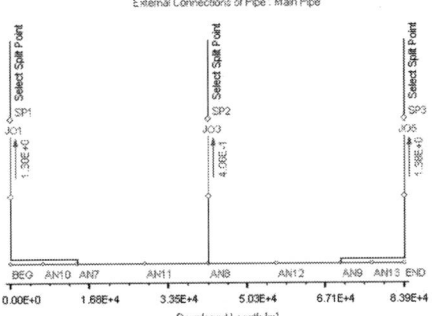

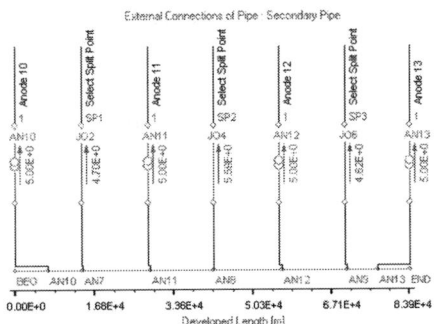

Figure 19: External connections for the main pipeline.

Figure 20: External connections for the secondary pipeline.

Do remark that the main pipeline gets an extra protection near the position of the additional four groundbeds of the secondary pipeline. This is not surprising since both pipelines are only 20m separated from each other.

From fig. 21 it can be seen that the average 'off' potential is only slightly more negative than before (except near groundbed AN8). This is due to the fact that the additional current of 0.08A is less than 3% of the original 3A. Similarly, the decrease of 0.08A in rectifier current for the secondary pipeline only gives a minor decrease in overall protection level as can clearly be seen from the results presented in fig. 22. Remark that near groundbed AN8, the protection level for the secondary pipeline is slightly better than in the isolated case.

7.4 Using experimental soil resistivity values as input

In practical situations, the soil resistivity along the trajectory of the pipeline is measured at a given number of intervals. These experimental resistivity data can directly be used within the calculations. An overview of such a file is given in fig. 23. Only a reduced number of data, 'measured' at the groundbed locations has

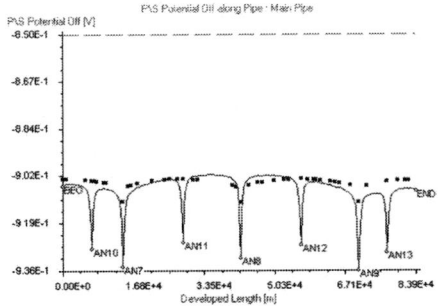

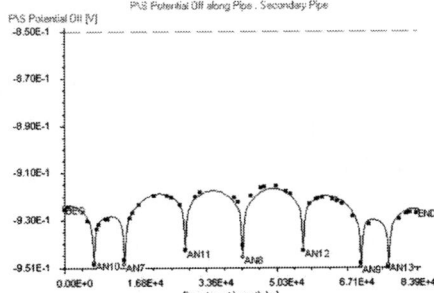

Figure 21: 'Off' potential along the main pipeline.

Figure 22: 'Off' potential along the secondary pipeline.

```
CatPro fieldfile version 1.3
2
7
L      RESIS
6966   20
14278  35
28457  50
42107  3
56602  5
70344  45
77102  60
```

Figure 23: Specifying the soil resistivity data for the main pipeline.

been entered. In between two successive data points, the soil resistivity is assumed to vary linearly.

The pipe-to-soil values along the main and secondary pipeline are presented in figs 24 and 25. The 'off' potential in the region between anodes 8 and 12 is more negative than before due to the reduced soil resistivity (ranging between 3 and 5Ω.m) and hence and increased current density. The regions near groundbeds AN11 and AN13, exhibiting the highest soil resistivity, are no longer protected. The secondary pipeline remains protected over the full length of it.

Finally, the coating resistance along both pipelines is shown in figs 26 and 27. For the main pipeline, the resistance ranges from 4780 to 93400Ω.m^2. The values for the secondary pipeline are about 20 times smaller.

8 Example 3: DC-traction stray currents influences

In this example it is explained how the software can be used to study DC-traction stray current influences on the cathodic protection of buried pipelines. This type of stray current is one of the most important sources of earth corrosion and generally results from the leakage of return currents from large DC-traction systems that are grounded or have a bad earth-insulated return path.

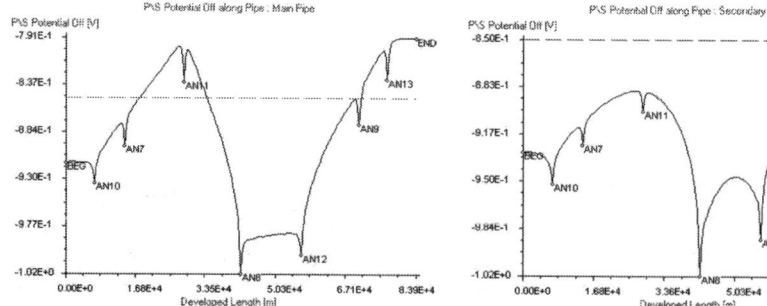

Figure 24: 'Off' potential for the main pipeline.

Figure 25: 'Off' potential for the secondary pipeline.

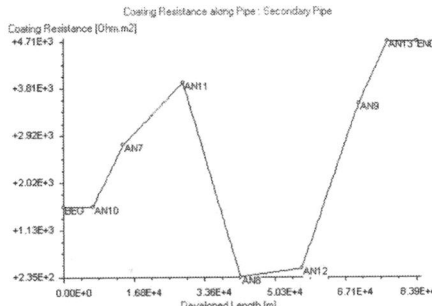

Figure 26: Coating resistance for the main pipeline.

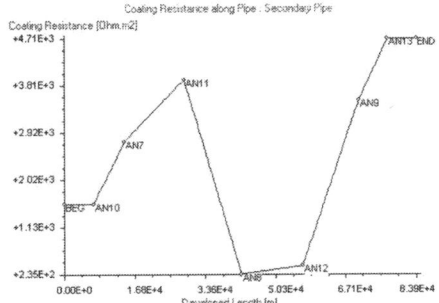

Figure 27: Coating resistance for the secondary pipeline.

The duration and the amplitude of the involved processes are important for the resulting metallic corrosion. Cathodic protection systems and railways rely for their function on large, persistent earth currents or suffer from periodic traffic, consuming substantial currents. Therefore the cathodic protection system becomes more complicated and has to be properly designed as will become clear from the calculations that follow.

8.1 Overview of the problem

Consider the practical situation of fig. 28. A large $30''$ gas transport pipeline with a total length of about 60km is protected by two deep-anode groundbeds. The centre of the pipe is located at a constant depth of 1.5m. The average soil resistivity along the pipeline trajectory is 100.0Ω.m. The groundbeds are at a depth of 20m and a distance of 50m with respect to the pipe.

On the left of the pipe, at a certain varying distance, lies a rail-road, composed out of two tracks, the first one crossing the pipeline. The power stations (sub-stations), indicated with PS1 to PS4, are at a distance of about 15km. It is evident that both tracks can be electrically connected in practice. This information is introduced by means of two joints JO1 and JO2, connecting both tracks and both overhead wires, respectively. Two trains (TR1,TR2), one on either track, are present. The polarisation behaviour for the pipe is described by the same non-linear polarisation curve as used for the old pipeline from the previous example (with a corrosion potential of -0.7V).

The coating quality is again described using a perfect coating (no holidays) with a number of distributed holidays. The holiday ratio is 0.05% and the average defect size is 1cm. Taking into account a soil resistivity of 100Ω.m, this yields a coating resistance of about 1610Ω.m^2.

Anode beds CS1 and CS2 are of the impressed current type and deliver respectively 3.9A and 4.1A (at 13.8V and 14.9V). This yields a resistance-to-ground of the anodes being 3.53Ω and 3.63Ω. The sub-stations PS1 to PS4 deliver a voltage of 1500V and the load resistance representing the traction current is 1Ω for both trains. The transition resistance between track and soil is $10,000\Omega$.m^2.

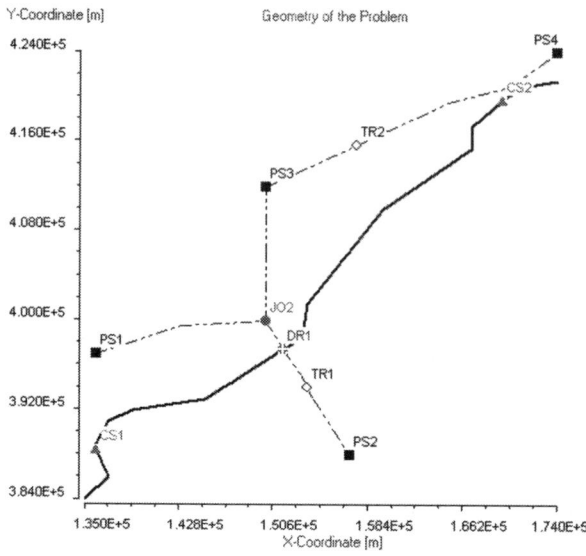

Figure 28: Simulation of DC-traction stray currents.

8.2 Simulation without trains

First, a simulation without trains is done. In fig. 29, the 'off' potential is plotted along the developed pipe length for the case with both anode beds operational. One can see that the entire pipeline is protected ('off' potential more negative than −0.85V versus CSE, see dashed red line) using only two anode beds.

A plot of the corresponding axial current along the pipeline is presented in fig. 30. This current is defined as being positive if it does flow in the direction from lower left to upper right on fig. 28. Negative values in fig. 30 therefore represent axial currents in the opposite direction. Note the two 'current jumps' at the position of the anode beds where respectively 3.9 and 4.1A are 'extracted' from the pipe.

8.3 Simulation with a train at position 1

Consider first train TR1 at position 1. The calculated 'off' potential is presented in fig. 31 and has been compared with the results from the stray current free simulation, presented in brown squares.

The train takes in total 1120A from the current feeder. This current is mainly delivered by sub-stations PS1 (172A), PS2 (733A) and PS3 (212A) and in less amount by PS4 (5.8A). The current returns to the sub-stations mainly via the rails. However, looking at the pipe-to-soil potential distribution along the pipeline in fig. 31, it appears that an important part of the current enters the pipe at the crossing with the rail. This effect induces locally a cathodic (over-)protection of the pipe that is even more important than the anode influences. This stray current reduces the local current density along some important parts of the pipeline.

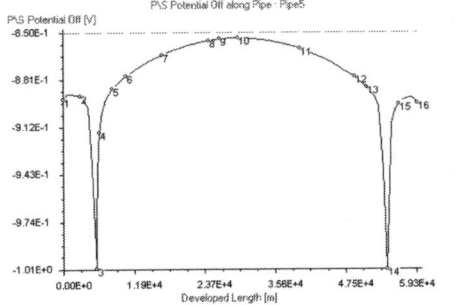 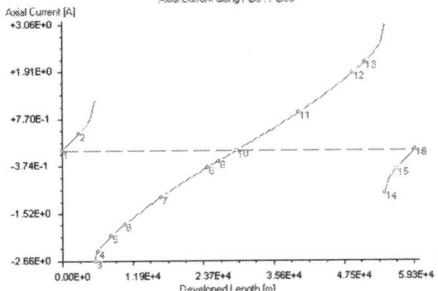

Figure 29: 'Off' potential (no trains).   Figure 30: Axial current (no trains).

Indeed, near location 11, the 'off' potential drops to only −0.84V, a little bit below the minimum protection level! In the region near anode bed CS1, the protection level is slightly lowered (but still more negative than −0.85V versus CSE) and the stray current flows via the rails back to sub-stations PS1 and (in less amount) PS2.

To the north however, the stray current picked up at the crossing severely reduces the protection level of the pipe. The rail between PS3 and PS4 plays an important role and it can be observed that large parts of the pipeline are just below the minimum protection level. Fortunately, but not by chance, the position of anode bed CS2 considerably reduces the effect.

From fig. 32, comparing the calculated 'on' and 'off' potential, it can be seen that near the crossing the stray current produces an 'on' potential that is about 950-mV more negative than the true 'off' potential! The IR-drop along the coating can be calculated from the current density profile and the coating resistance. It turns out that the IR-drop along the coating is 515mV. The other 435mV potential drop is due to the soil between the pipe and the reference electrode at the earth surface.

8.3.1 Potentials along track and overhead wire

The stray current pattern generated by train TR1 strongly depends on the potential of the track, as shown in fig. 33.

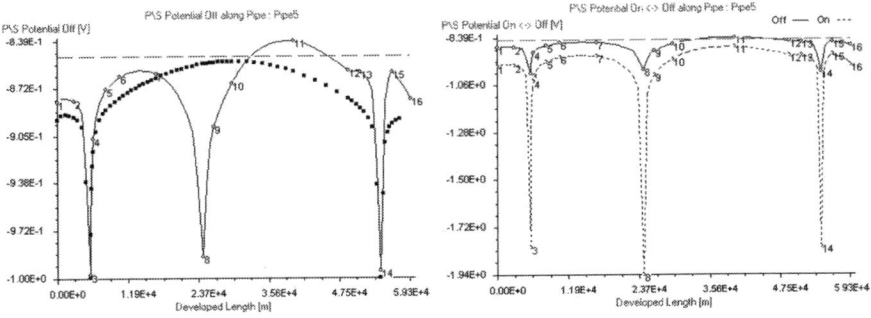

Figure 31: 'Off' potential (TR1 active).   Figure 32: 'On' and 'off' potential (TR1 active).

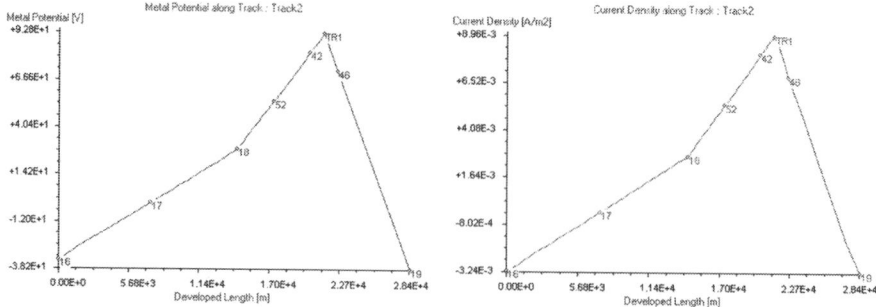

Figure 33: Track potential (TR1 active).   Figure 34: Track current density (TR1 active).

From this plot is can be seen that the highest track potential (+92.8V) is found at the position of the train, and the lowest one near the sub-stations PS1 (−32.5V) and PS2 (−38.2V). Track regions at high voltages generate stray currents that leave the rail (positive current density) and return back to the track (negative current density) near the sub-stations as can be seen from the rail current density plot of fig. 34.

8.3.2 External network belonging to a track

A complete overview of the track-wire network with all external connections (trains, sub-stations, drains, joints, ...) is shown in fig. 35. From this figure it can be seen that the train indeed receives 1120A and that 182A flows to the other track through joint JO1. The overhead wire receives 217A, injected via joint JO2.

8.4 Simulation with a train at position 2

Next consider train TR2 at position 2 (TR1 being disabled). The train takes 1050A from the current feeder. This current is almost completely delivered by sub-stations PS3 (721A) and PS4 (321A) and returns mainly to the sub-stations via the rails. The corresponding pipe-to-soil potential along the pipeline is given in fig. 36.

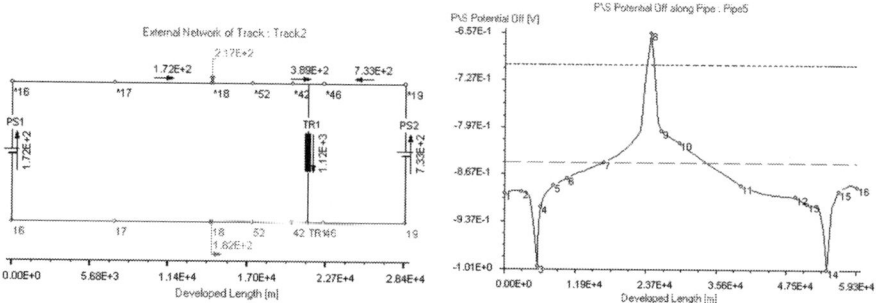

Figure 35: External   network   (TR1   Figure 36: 'Off' potential (TR2 active).
            active).

An important part of the traction current leaves the pipe at the crossing with the rail and returns via the rail back to sub-station PS3. As a result the part of the pipe near the crossing is made anodic (pipe-to-soil potential higher than corrosion potential) and severe local corrosion occurs. The stray current enters the pipeline in the neighbourhood of anode bed CS2 where the distance between rail and pipe is small. The effect of the stray current is added to the effect of the anode bed CS2. Overprotection might take place. The region near to anode bed CS1 is scarcely influenced during this situation.

It is interesting to have a closer look at fig. 37, comparing the calculated 'on' and 'off' potential. One can clearly observe that in a zone near the crossing, the calculated 'on' potential is indeed less negative than the corresponding 'off' potential, clearly indicating that at this position, current leaves the pipeline and corrosion occurs. This is also confirmed by the current density profile presented in fig. 38.

8.5 Influence of a current drain at the crossing

When the sources of stray currents are accessible, current drainage techniques can be implemented. This means that the structure that suffers from stray current corrosion, being the pipe, is connected to the negative pole of the DC-source that generates the stray current (*i.e.* the first track). This metallic connection must have a lower resistance than the alternative earth return path.

Moreover, the current drain is made unidirectional (*i.e.* a diode in series with a resistance R), such that current can only flow from the pipe to the track. Stray currents that entered the pipe will then prefer this connection to return to the generator, instead of the soil. The amount of current that is drained from the pipeline can be controlled by changing the resistance R of the drainage.

In practice, such a unidirectional drainage is present at the crossing of the pipe and the track as can be seen from fig. 28 (DR 1). Simulations with R respectively equal to 10.0, 5.0 and 2.0Ω have been performed, resulting in a drain current of 2.0, 3.8 and 8.6A, respectively. From fig. 39 ($R = 5.0\Omega$) and fig. 40 ($R = 2.0\Omega$) it is clearly noticed that the drainage with the lowest resistance is working well,

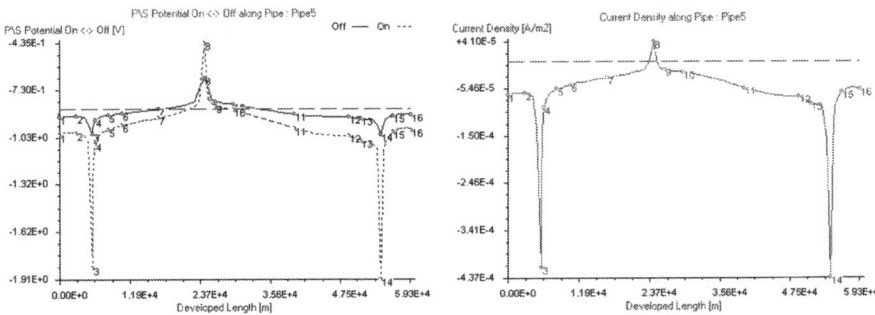

Figure 37: 'On' and 'off' potential (TR2 active).

Figure 38: Current density (TR2 active).

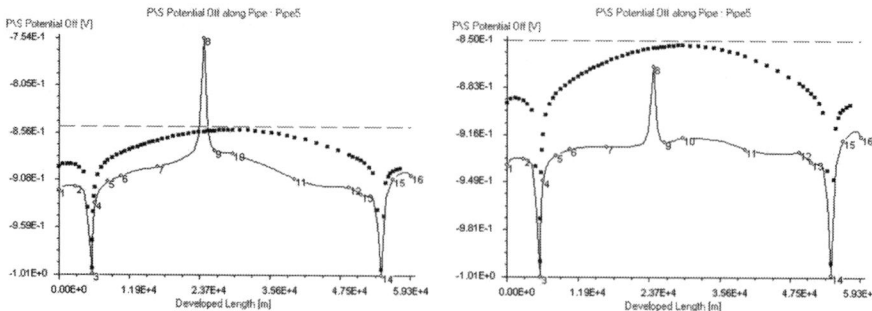

Figure 39: 'Off' potential (TR2 active drain resistance = 5Ω).

Figure 40: 'Off' potential (TR2 active drain resistance = 2Ω).

ensuring a protection of the pipeline along the whole length (squares represent stray current free simulations). Although not presented here, it turns out that a drain with negligible resistance R gives a protection level near the crossing of -1.04V, comparable to that near the groundbeds. The drain current in this case is about 53.5A.

8.6 Influence of the joint between the tracks and wires

As mentioned in the introduction of the problem, both tracks and overhead wires are connected to each other by means of joints JO1 and JO2, respectively. When both joints are active, the external networks associated with both tracks are, from an electrical point of view, fully coupled. In what follows, it will be investigated how the stray currents behave if one or both of the joints are inactive.

Let's first find out what happens to the stray current pattern produced by TR1 at position 1. Therefore, in total 4 calculations have been done with JO1 and JO2 switched on and off. An overview of the results found is presented in table 1. For each run, the currents through the joints (when active) and the sub-stations have been presented, followed by the pipe-to-soil potential (protection level) of the pipe at the crossing and at the end of the pipeline (point 15, just after anode bed CS2).

Joint JO1 has been defined from the first to second track. This means that a negative value in table 1 indicates a current that flows in the opposite direction. The same remark holds for joint JO2.

Table 1: Influence of JO1 and JO2 on the current balances (TR1 active).

case	JO1 [A]	JO2 [A]	PS1 [A]	PS2 [A]	PS3 [A]	PS4 [A]	PSP (8) [V]	PSP (15) [V]
on-on	182	−217	172	733	212	5.8	−0.99	−0.86
on-off	18.7	N.A.	268	813	1.8	−1.8	−0.96	−0.89
off-on	N.A.	−128	210	765	115	13.7	−1.03	−0.85
off-off	N.A.	N.A.	268	813	0.01	−0.01	−0.97	−0.89

From this table and the overview fig. 41, it can clearly be seen that the largest stray current at the crossing and the lowest protection level at the end of the line are obtained when JO2 is active and JO1 is inactive. In this case, the train still receives current from PS3 and PS4 since JO2 is active, explaining the large protection level (stray current pick-up) at the crossing. However, the return path for the traction current is partly blocked since JO1 is inactive, encouraging the current to go back to PS3 and PS4 via the pipe, and hence explaining the reduced protection level of the pipe between both sub-stations.

In fig. 42, all external connections associated with the second track are given in the case when JO1 is active and JO2 is not. From this figure and table 1 it can be seen that a current of 18.7A originates from the first track and enters the network through JO1. Part of this current (1.78A) goes through PS3 and circulates in the network between both sub-stations, other parts follow the track and leak into the soil. Remark that the metal potential of the track at PS3 and PS4 is 8.3V and 6.9V respectively with a minimum of 5.9V at the point nearest to the pipeline in the region of CS2.

A summary of the stray current patterns produced by TR2 at position 2 (TR1 and DR 1 are inactive) is presented in table 2. From this table and the overview fig. 43, it can be remarked that the influence of JO1 in this situation is almost completely negligible. When both joints are active the current through JO1 is only 25.4A compared to the 182A in the previous case (TR1 at position 1). Again, the largest stray current influences at both the crossing and at the end of the pipeline are obtained when JO1 and JO2 are both active.

8.7 Influence of unbalanced sub-station voltages

In the previous calculations, the influence of traction current leaks on the cathodic protection of the pipeline has been investigated. However, stray currents can also originate from DC-traction systems when no trains are present due to unbalanced

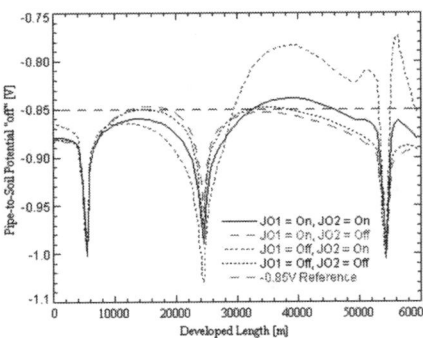

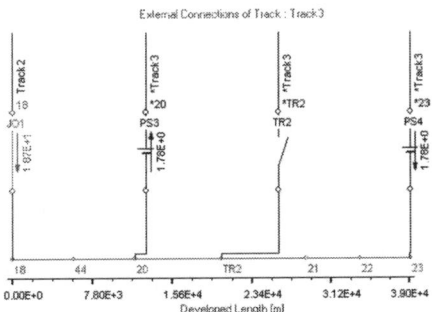

Figure 41: 'Off' potential for different modes of JO1 and JO2 (TR1 active).

Figure 42: External connections associated with the second track (TR1, JO1 active; JO2 inactive).

Table 2: Influence of JO1 and JO2 on the current balances (TR2 active).

case	JO1 [A]	JO2 [A]	PS1 [A]	PS2 [A]	PS3 [A]	PS4 [A]	PSP (8) [V]	PSP (15) [V]
on-on	25.4	12.9	6.4	6.5	721	321	−0.66	−0.89
on-off	35.1	N.A.	0.1	−0.1	733	321	−0.66	−0.89
off-on	N.A.	25.1	12.3	12.7	708	322	−0.69	−0.88
off-off	N.A.	N.A.	0	0	729	324	−0.83	−0.88

sub-station voltages. To demonstrate this the voltage delivered by PS1 and PS2 will be increased with 5% and compared with the standard situation were all sub-stations are at 1500V. For each test case, the metal potential of the track at PS1 and PS2 has been noted. At the crossing, the protection level ('off' potential) of the pipe, the metal potential of the pipe and the metal potential of the track have been investigated.

An overview of the results is given in table 3. From this table it can be seen that when all sub-stations are perfectly balanced, the metal potential of the track at the crossing is very close to zero. However, when one of the sub-stations is out of balance, a continuous compensation current will flow between PS1 and PS2. The track potential at the sub-station that delivers the highest voltage will be the lowest, *i.e.* −18.5V in both cases. The worst case however is encountered when PS2 (being closer to the crossing than PS1) delivers the highest voltage (1575V). In this case, the metal potential of the track at the crossing with the pipe is the lowest (*i.e.* −5.35V), which will increase the discharge of current from the pipe at that position. Therefore the protection level of the pipe at the crossing drops to −0.78V, which is below the minimum value. An overview of the protection level of the whole pipeline in this situation is given in fig. 44.

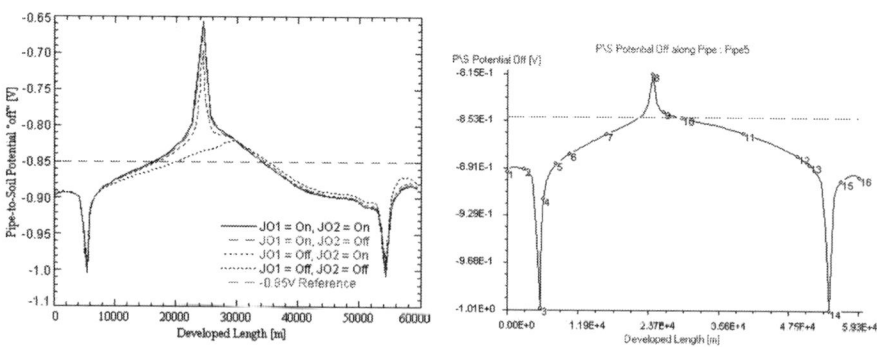

Figure 43: 'Off' potential for different operation modes of JO1 and JO2 (TR2 active).

Figure 44: 'Off' potential (PS2 at 1575V).

Table 3: Influence of unbalanced voltages between PS1 and PS2 (no trains).

PS1 [V]	PS2 [V]	$V_{t,PS1}$ [V]	$V_{t,PS2}$ [V]	$V_{p,cross}$ [V]	$V_{t,cross}$ [V]	PSP_{cross} [V]
1500	1500	+0.0	+0.0	−0.87	+0.0	−0.86
1575	1500	−18.5	+5.8	−0.87	−0.24	−0.85
1500	1575	+5.8	−18.5	−0.87	−5.34	−0.78

9 Conclusion

A simulation software dedicated to the modelling of the cathodic protection of networks of buried pipeline has been presented. This model accurately deals with problems involving multiple cathodic protection systems. As a result, interference situations occurring when several CP-systems interact, can be calculated. It has been shown how the coupling with an additional aboveground electrical network and the introduction of special pipes, representing railways, allows to model stray currents arising from traction systems.

The extended model has been tested on underground corrosion situations involving parallel pipelines and traction systems. The dc-traction calculation clearly shows how stray current can have important influences at a long distance from their place of creation. Their impact on the overall protection level of the pipeline depends on several parameters such as the overall geometry, train positions, sub-station voltages, joints, ... The combined effect of all those influences makes it very difficult or even impossible to predict the outcome without a powerful simulation tool that can investigate all these influences. It has been proven that the developed model is a powerful, flexible and straightforward simulation tool that enables calculation of many real situations encountered in corrosion of underground pipelines.

References

[1] Brichau, F. & Deconinck, J., A numerical model for cathodic protection of buried pipes. *CORROSION*, **50(1)**, pp. 39–49, 1994.

[2] Brichau, F., Deconinck, J. & Driesens, T., Modeling of underground cathodic protection stray currents. *CORROSION*, **52(6)**, pp. 480–488, 1996.

[3] Catpro software manual, www.elsyca.com.

[4] Brebbia, C.A., Telles, J. & Wrobel, L., (eds.) *Boundary Element Techniques*. Springer-Verlag: Berlin and New York, 1984.

[5] Orazem, M., Esteban, J., Kennelley, K. & Degerstedt, R., Mathematical model for cathodic protection of an underground pipeline with coating holidays: Part 1 – theoretical development. *CORROSION*, **53(4)**, pp. 264–272, 1997.

[6] Orazem, M., Esteban, J., Kennelley, K. & Degerstedt, R., Mathematical model for cathodic protection of an underground pipeline with coating holidays: Part 2 – case studies of parallel anode cathodic protection systems. *CORROSION*, **53(6)**, pp. 427–436, 1997.

[7] Biase, L.D., Corrosione da corrente alternata su tubazioni metallische interrate: stato dell'arte e prospettive. *Convegno Nazionale APCE Roma*, 1996.

[8] Modjtahedi, D. & Jamali, M., BEM with fundamental sources placed on an imaginary boundary. *Proceedings of 12th International Conference on Boundary Elements Method*, pp. 111–121, 1990.

[9] Carlson, L., Fitzgerald, J. & Webster, R., Cathodic protection design for 1, 900 miles (3050 km) of high-pressure natural gas pipeline. *Materials Performance*, **40(8)**, pp. 28–32, 2001.

Modeling coating flaws with non-linear polarization curves for long pipelines

D.P. Riemer[1] & M.E. Orazem[2]
[1] *OLI Systems, Inc., USA.*
[2] *Department of Chemical Engineering, University of Florida, USA.*

Abstract

External corrosion is commonly mitigated by coating the structure with a high resistance film and by employing cathodic protection (CP) to protect regions that are inadequately coated or where the coating has degraded. Defects in the coating are termed holidays, and such holidays can expose bare steel. The objective of this paper is to explore the state-of-the-art in computer models for cathodic protection. The emphasis is placed here on mitigation of corrosion of underground pipelines, but the concepts, models, and techniques described are sufficiently general to be applied to mitigation of corrosion of any structure.

1 Introduction

Since the beginning of the 20th century, petroleum products and natural gas have been transported over long distances by buried steel pipelines. Over 1.3 million miles of buried steel main-line pipe are used to transport natural gas within the United States [1]. An additional 170,000 miles of pipeline are used to transport crude oil and refined petroleum products [2].

The limited availability of right-of-way corridors requires that new pipelines be located next to existing pipelines. Placement of pipelines in close proximity introduces the potential for interference between systems providing cathodic protection to the respective pipelines. In addition, the modern use of coatings, introduced to lower the current requirement for cathodic protection of pipelines, introduces as well the potential for localized failure of pipes at discrete coating defects. The prediction of the performance of cathodic protection systems under these conditions requires a mathematical model that can account for current and potential distributions in both angular and axial directions.

The objective of this work is to describe a mathematical model that has been developed for cathodic protection of an arbitrary number of pipelines linked to an arbitrary number of cathodic protection systems. The Boundary Element Method was used in the model formulation. In order to achieve the desired calculation accuracy and speed, the development required significant refinement to algorithms available in the literature.

2 Mathematical development

The model for cathodic protection of pipelines must account for the flow of current in the soil, in the pipes, and in the circuitry. Until recently, most models of cathodic protection of pipelines assumed that the potential of the pipe steel was uniform. The assumption of a uniform steel potential enabled solution for current and potential distributions without consideration of the passage of current through the pipe. Long pipes, however, exhibit a non-negligible potential difference along the steel [3–5].

In the context of the present work, there exist two separate domains for the flow of current. The first is the soil (or outer) domain, bounded by the surfaces of the pipes and anodes, the interface between the soil and air, and, if present, the interface between the soil and any buried insulating layers. The second (inner) domain comprises the metallic wall of the pipe, the volume of the anode, and connecting wires and resistors for the return path of the protective current. The two domains are linked by the electrochemical reactions described in section 2.3.

2.1 Soil (or outer) domain

The soil domain comprises the material in which the pipes and anodes are buried. From the perspective of the pipe, this domain can be considered to be an outer domain. This designation contrasts the soil domain from the domain which includes the interior volume of pipeline steel.

In principle, the treatment of concentrations and potential within the soil requires solution of a coupled set of equations, including conservation of each individual solute species [6], *i.e.*

$$\frac{\partial c_i}{\partial t} = -(\nabla \cdot \mathbf{N}_i) + R_i \tag{1}$$

and an expression of electroneutrality

$$\sum_i z_i c_i = 0 \tag{2}$$

where c_i is the concentration of species i, R_i is the rate of generation of species i due to homogeneous reactions, and $\mathbf{N}_i$ is the net flux vector for species i. In a dilute electrolytic solution, the flux N_i includes contributions from convection, diffusion, and migration as

$$N_i = vc - D_i \left[\nabla c_i - \frac{z_i c_i F}{RT} \nabla \Phi \right] \tag{3}$$

where v is the fluid velocity, D_i is the diffusion coefficient for species i, z_i is the charge associated with species i, F is Faraday's constant, and Φ is the potential.

Under the assumption of a steady-state and a uniform concentration of ionic species, the current density, expressed in terms of contributions from the motion of each ionic species, can be given in terms of potential by Ohm's law. Thus,

$$i = F \sum_i z_i N_i = -\kappa \nabla \Phi \tag{4}$$

where the conductivity κ, expressed in terms of contributions from individual species as

$$\kappa = F^2 \sum_i z_i^2 u_i c_i \tag{5}$$

has a uniform value because the concentration is uniform. The assumption that concentrations are uniform yields

$$\nabla^2 \Phi = 0. \tag{6}$$

Thus, potential is governed by Laplace's equation.

Equation (6) is commonly used without explanation in cathodic protection models. The development presented above can be used to emphasize some important points.

- The assumption that concentrations are uniform means that concentration gradients associated with reactions on the pipe and anode surface are assumed to occur within a thin layer adjacent to the pipe and anode surfaces. The concentration gradients within this narrow layer can be incorporated into the boundary condition which describes the electrochemical reactions.
- While changes in soil resistivity can be treated within Boundary Element models by creating coupled soil domains, the variation of potential across the boundaries between domains must include the contribution of a diffusion potential associated with concentration gradients.

The resistivity of the soil domain was assumed to be uniform in the present model.

2.2 Pipe metal (or inner) domain

The potential drop within the pipeline steel can be very significant when the current level is large and the pipelines are long. The Ohmic resistance of a ANSI Standard 18-inch (0.457m) diameter steel pipe with a wall thickness of 0.375 inches (0.95cm) is 7.2×10^{-3} Ω/km. The potential drop along a 100km length of this pipe would be 720mV for passage of a relatively modest 1A of CP current. The assumption that the potential drop in the pipe steel can be neglected, made, for example, by Esteban *et al* [7] for their treatment of coating holidays in a short length of pipe, cannot be made for longer stretches of pipeline networks.

The flow of current through the pipe steel, anodes, and connecting wires is strictly governed by the Laplace's equation

$$\nabla \cdot (\kappa \nabla V) = 0 \tag{7}$$

where V is the departure of the potential of the metal from a uniform value and κ is the material conductivity. The conductivity of the pipe-metal domain κ is not necessarily uniform. For example, the copper wires connecting the pipe to the anode have a different conductivity than does pipeline steel or the anode, or groundbed material. A simplified version of eqn (7)

$$\Delta V = IR = I\rho \frac{L}{A} \tag{8}$$

can be used to account for the potential drop across connecting wires, where R is the resistance of the wire, ρ is the electrical resistivity, L is the length, and A is the cross-sectional area of wire.

2.3 Domain coupling through boundary conditions

The inner and outer domains described in sections 2.1 and 2.2, respectively, are linked by boundary conditions which relate the current density on metal surfaces to values of local potential. The type of surface governs the specific form of the relationship required. Bare steel, coated steel, galvanic anodes, and impressed current anodes are considered in the following sections.

2.3.1 Bare steel

While the surface of modern pipelines are generally covered by a protective coating, bare metal can be exposed by coating defects caused by mechanical damage or long-term degradation mechanisms. For steel in soil, the electrochemical reactions include corrosion, oxygen reduction, and, at sufficiently cathodic potentials, hydrogen evolution. One common form of the boundary condition, adapted for steady-state soil systems by Yan *et al* [8] and Kennelly *et al* [9], takes the form

$$i = 10^{(V-\Phi-E_{Fe})/\beta_{Fe}} - \left(\frac{1}{i_{\lim,O_2}} - 10^{(V-\Phi-E_{O_2})/\beta_{O_2}} \right)^{-1}$$
$$- 10^{-(V-\Phi-E_{H_2})/\beta_{H_2}} \tag{9}$$

where V is the potential of the steel, obtained from solution of the inner domain, and Φ is the potential of the soil adjacent to the steel, obtained from solution of the outer domain. The term β_{Fe} represents the Tafel slope for the corrosion reaction, and E_{Fe} represents the equilibrium potential for the corrosion reaction. Similar terms are used for the oxygen reduction and hydrogen evolution reactions.

The mass transfer of oxygen is taken into account by including the mass transfer limited current density for oxygen reduction $i_{\lim,O_2}$.

The influence of calcareous and corrosion-product films on the bare surface can be taken into account by adjusting the kinetic parameter values. Nisançioglu [10–12] showed how the parameters of polarization curves can be adjusted in CP models to account for the influence of polarization history in transient models. Carson and Orazem suggested a refined regression approach to obtain time-dependent polarization parameters [13]. Polarization curves can be obtained for specific chemical compositions through predictive mathematical models such as that given by Anderko *et al* [14, 15]. The principle is that the current contributions for individual reactions are summed to yield a relationship between local current density, the potential of the steel, and the potential of the soil adjacent to the steel.

2.3.2 Coated steel

Application of coatings to steel before burial reduces the amount of current required to provide cathodic protection. These coatings are usually polymeric but can sometimes be cement-based products.

Cathodic protection must nevertheless be used for coated pipes. Isolated faults in the coating (termed holidays) can form galvanic couples with the coated portion, thereby accelerating the corrosion of the bare steel exposed by the holiday. The pipeline will experience failure due to corrosion in a far shorter time than would be experienced if the pipe were left bare [3].

The coated portion of steel is often treated as an insulator or as a simple high-resistance electronic conductor. To explore the role of galvanic coupling, however, the electrochemical reactions on the coated steel must be treated explicitly. It has been reported that a small current will pass through some coatings after they have absorbed sufficient amounts of water [16–20]. Corfias *et al* have shown that the pore structure expands as the coating absorbs water [17]. They have also shown that the coating conductivity increases significantly with immersion time.

Two possible modes for transport through a coating are illustrated schematically in fig.1. Figure 1(a) represents a uniform barrier in which ionic species enter a polymeric phase. The electrochemical reaction is assumed to take place at the metal-coating interface, driven by the potential difference $V - \Phi_{in}$. Figure 1(b) represents a metal covered by an insulating coating perforated by pores through which ionic

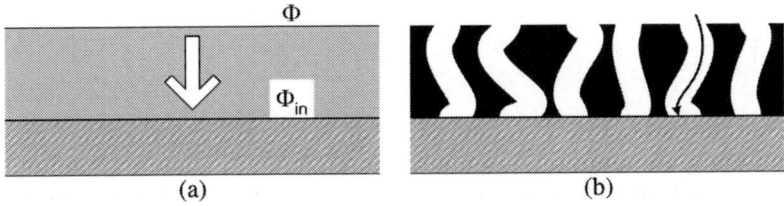

Figure 1: Schematic representation of two models for coating behavior: a) coating allowing uniformly distributed transport of species, and b) coating with discrete pores through which transport can take place.

species may move. The effective resistivity of the coating then depends on the number of pores per unit area. Again, the electrochemical reaction is assumed to take place at the metal-coating interface and is driven by the potential difference $V - \Phi_{\text{in}}$. The view of a coated steel represented in fig. 1 is supported by results reported by NOVA Gas [21] and CC Technologies [19, 22, 23], which showed that coatings on steel pipe formed a diffusion barrier when placed in aqueous environments, that the coating absorbs water, and that it is possible to polarize slightly the steel under a disbonded coating.

The modification of eqn (9) by Riemer and Orazem [24, 25] can address either mode of transport illustrated in fig. 1. The potential drop through the film or coating can be expressed as [7]

$$i = \frac{\Phi - \Phi_{\text{in}}}{\rho \delta} \tag{10}$$

where Φ is the potential in the electrolyte next to the coating, Φ_{in} is the potential at the underside of the coating just above the steel, ρ is the effective resistivity of the coating and δ is the thickness of the coating. The current density can also be written in terms of the electrochemical reactions as

$$i = \frac{A_{\text{pore}}}{A} \left[10^{(V - \Phi_{in} - E_{Fe})/\beta_{Fe}} - \left(\frac{1}{(1 - \alpha_{blk}) i_{lim,O_2}} - 10^{(V - \Phi_{in} - E_{O_2})/\beta_{O_2}} \right)^{-1} \right.$$
$$\left. - 10^{-(V - \Phi_{in} - E_{H_2})/\beta_{H_2}} \right] \tag{11}$$

where A_{pore}/A is the effective surface area available for reactions, and α_{blk} is the reduction to the transport of oxygen through the barrier. It is also assumed that the coating has absorbed enough water that the hydrogen evolution reaction is not mass-transfer limited. Equations (10) and (11) are solved simultaneously by a Newton–Ralphson method to get values for the current density and Φ_{in}.

By eliminating the current density in eqns (10) and (11), a particularly useful equation is obtained as

$$\frac{A \left(\Phi - \Phi_{\text{in}} \right)}{A_{\text{pore}} \rho_{\text{film}} \delta_{\text{film}}} = 10^{(V - \Phi_{\text{in}} - E_{\text{Fe}})/\beta_{\text{Fe}}}$$
$$- \left(\frac{1}{(1 - \alpha_{\text{blk}}) i_{\text{lim},O_2}} - 10^{(V - \Phi_{\text{in}} - E_{O_2})/\beta_{O_2}} \right)^{-1} - 10^{-(V - \Phi_{\text{in}} - E_{H_2})/\beta_{H_2}} \tag{12}$$

which links V, Φ, and Φ_{in}. Equation (12) can be solved by a Newton–Ralphson method to get Φ_{in} for given values of V and Φ. The current density can subsequently be calculated from eqn (10).

The corrosion potential of the coated steel can be calculated from eqn (11) under the assumption that no current flows through the coating. If $(1 - \alpha_{\text{blk}})$ is smaller than the pore ratio, then the metal will be more active than bare steel. If they are the same, then the coated pipe and bare steel will have the same corrosion potential. If $(1 - \alpha_{blk})$ is larger than the pore ratio, the metal under the coating will be more noble than the bare steel and will contribute to a galvanic couple that, in the absence of CP, will enhance the corrosion rate at the holiday.

2.3.3 Galvanic anodes

The treatment of electrochemical processes at galvanic or impressed-current anodes is similar to that used in section 2.3.1 for bare steel. A galvanic anode is selected to be anodic to the metal to be protected. Zinc and magnesium are used to protect carbon steel in soil environments, and aluminum and magnesium are used in sea water. The aluminum–steel couple does not provide a sufficient driving force to be effective in soil where the resistivity is typically much higher than that of sea water.

The electrochemical reactions at the galvanic anode typically consist of oxygen reduction, assumed to take place at the mass-transfer-limited rate, and corrosion. The resulting expression is given as

$$i = i_{O_2}\left(10^{(V - \Phi - E_{corr})/\beta_{anode}} - 1\right) \tag{13}$$

where i_{O_2} is the mass-transfer-limited current density for oxygen reduction, E_{corr} is the free corrosion (equilibrium) potential of the anode, and β is the Tafel slope for the anode corrosion reaction. The assumption that oxygen reduction is mass-transfer-controlled is justified at the potentials typically seen for galvanic anodes. Hydrogen evolution was ignored for galvanic anodes because it only makes a small contribution to the net current at operating potentials. Some typical values reported in the literature for the model parameters used in eqn (13) are given in table 1. The value for the mass-transfer-limited current density must be obtained for specific soil conditions.

2.3.4 Impressed-current anodes

Impressed-current systems consist of a dimensionally stable (non-corroding) anode connected to the positive terminal of a direct-current (DC) rectifier. The negative terminal of the rectifier is connected to the pipe. The rectifier can then be used to push the potential of the anode to any desired potential that is more negative than that

Table 1: Parameters for common galvanic anodes [4].

Anode Type	E_{eql} (mV CSE)	β (mV/decade)	i_{O_2} ($\mu A/cm^2$)
Al	-1.0	60	1.0
Zn	-1.1	60	1.0
Mg (Standard)	-1.5	60	1.0
Mg (High Potential)	-1.75	60	1.0

Table 2: Parameters for the oxygen and chlorine evolution reactions [26].

Reaction	Equilibrium Potential (E)	Tafel Slope (β)
O_2 evolution	-172mV (CSE)	100mV/decade
Cl_2 evolution	50mV (CSE)	100mV/decade

of the pipe. Because the driving potential for impressed-current anodes is provided by an outside current source, the current densities obtainable on impressed-current systems are usually an order of magnitude larger than can be obtained with galvanic anodes. Impressed-current systems can therefore protect a much longer section of pipe and can be used to greater effect in highly resistive soils.

The anodic reaction at an impressed-current anode is typically water oxidation (evolution of oxygen)

$$2H_2O \rightarrow O_2 + 4H^+ + 4e^- \tag{14}$$

and, at more anodic potentials, chloride oxidation (evolution of chlorine)

$$2Cl^- \rightarrow Cl_2 + 2e^-. \tag{15}$$

The cathodic reaction would be the reduction of oxygen, just as was seen in eqn (13) for galvanic anodes. Thus, the polarization model for an impressed-current anode resembles that for galvanic anodes with the exception that an additional term is added to account for the potential setting of the rectifier, *i.e.*

$$i = i_{O_2} \left(10^{\left(V - \Phi - \triangle V_{\text{rect}} - E_{O_2} \right)/\beta_{O_2}} - 1 \right) \tag{16}$$

where $\triangle V_{\text{rect}}$ is the potential added by the rectifier, V is the voltage of the anode, Φ is the voltage just outside the surface of the anode, E_{O_2} is the equilibrium potential for the oxygen evolution reaction, and β_{O_2} is the tafel slope for the oxygen evolution reaction.

If chloride ions are present, eqn (16) must be modified to account for the contribution of the chlorine evolution reaction (15). Reaction (15) takes place in environments such as salt marshes and estuaries which can contain significant amounts of salt.

Kinetic parameters for reactions (14) and (15) are given in table 2. Again, the mass-transfer-limited current density for oxygen reduction i_{O_2} must be determined experimentally for a specific location.

3 Numerical development

Laplace's equation has been solved for many boundary conditions and domains [27]. Analytic and semi-analytic formulations can be used for specific geometries. For example, Moulton first applied the Schwartz-Christoffel transformation to calculate analytically the conduction of current through a two-dimensional rectangular geometry with arbitrarily placed electrodes [28]. Bowman supplies details for analytic solutions using the Schwartz-Christoffel transformation [29]. Orazem and Newman provide a semi-analytical implementation of the Schwartz-Christoffel transformation for the more complicated structure of a slotted electrode [30]. Orazem used the technique for a compact tension fracture specimen [31] which was later further refined [32]. Diem *et al* take the semi-analytical technique a step further by allowing for insulating surfaces to form an arbitrary angle with the electrode [33].

Unfortunately, such analytic and semi-analytic approaches are not sufficiently general to allow all the possible configurations of pipes within a domain, the detailed treatment of potential variation within the pipes, and the polarization behavior of the metal surfaces. Thus, numerical techniques are required. Of the available techniques, the boundary element method is particularly attractive because it can provide accurate calculations for arbitrary geometry. The method only solves the governing equation on the boundaries, which is ideal for corrosion problems where all the activity takes place at the boundaries. Brebbia first applied the boundary element method for potential problems governed by Laplace's equation [34]. Aoki [35] and Telles [36] reported the first practical utilization of the boundary element method with simple nonlinear boundary conditions. Zamani and Chuang demonstrated optimization of cathodic current through adjustment of anode location [37].

The pipe steel domain is solved using the finite element method. Brichau first demonstrated the technique of coupling a finite element solution for pipe steel to a boundary element solution for the soil [38]. He also demonstrated stray current effects from electric railroad interference utilizing the same solution formulation [39]. However, Brichau's method was limited in that it assumed that the potential and current distributions on the pipes and anodes were axisymmetric allowing only axial variations. Aoki presented a technique similar to Brichau's that included optimization of anode locations and several soil conductivity changes for the case of a single pipe with no angular variations in potential and current distributions [40, 41].

Since it is desired to have a solution for the current and potential distributions both around the circumference and along the length of the pipe, Brichau's method must be modified. These modifications are described in the following sections. The elements of the BEM and FEM techniques which are well established in the literature are summarized here for completeness.

3.1 Outer domain: boundary element method

The boundary element method can be derived from the same technique used to obtain the classical finite element method. One starts by writing a variational or weighted residual of the governing differential equation. If the PDE takes the form

$$\nabla^2 u - f = 0 \tag{17}$$

where f is a forcing term that is a function of position only, then one would write the weighted residual as

$$\int_\Omega w\left(\nabla^2 u - f\right) d\Omega = 0, \forall w \tag{18}$$

where Ω is the domain and w is any weighting function. In the case of Laplace's equation, $f = 0$ and $u = \Phi$; thus

$$\int_\Omega w\nabla^2\Phi d\Omega = 0, \forall w. \tag{19}$$

This equation holds for all weighting functions w. Equation (19) is integrated analytically by using the divergence theorem

$$\int_\Omega \nabla w\nabla\Phi d\Omega - \int_\Gamma w\left(\vec{n}\cdot\nabla\Phi\right) d\Gamma = 0, \forall w \tag{20}$$

with Γ being the boundary of the domain. Equation (20) is the classical weak form of the finite element method for Laplace's equation. At this stage, the boundary element method development departs from the finite element method by using a second application of the divergence theorem. The highest order derivative is thereby moved to the weighting function, *i.e.*

$$\int_\Omega \Phi\nabla^2 w d\Omega - \int_\Gamma w\left(\vec{n}\cdot\nabla\Phi\right) d\Gamma + \int_\Gamma \Phi\left(\vec{n}\cdot\nabla w\right) d\Gamma = 0, \forall w. \tag{21}$$

At this point the weighting function needs to be specified to show why the second application of the divergence theorem was done. If one observes that the solution to the equation

$$\nabla^2 G_{i,j} + \delta_i = 0 \tag{22}$$

is the Greens function for Laplace's equation, one can simplify eqn (21) by picking the weighting function to be the Green's function [42]. The first integral in eqn (21) becomes

$$\int_\Omega \Phi\left(-\delta_i\right) d\Omega = -\Phi_i. \tag{23}$$

Substitution of eqn (23) into eqn (21) yields a simpler equation valid within the domain

$$\Phi_i + \int_\Gamma \Phi\left(\vec{n}\cdot\nabla G_{i,j}\right) d\Gamma = \int_\Gamma G_{i,j}\left(\vec{n}\cdot\nabla\Phi\right) d\Gamma \tag{24}$$

where the highest order derivative has been removed and all the integrals are along the boundaries only. Equation (24) is still exact in so far as the Green's function is known and is valid for determining the values of the potential at any interior point in the domain given that the potential and current density distributions on the boundary are known. This implies that, for Laplace's equation, values of interior points are fully specified by integrals along the boundary only.

The last step in deriving the Boundary Element method is to take Φ_i to the boundary. A Cauchy principle value is introduced in the integral on the left side of eqn (24). It is usually represented by a constant appearing in front of the first term in eqn (24)

$$c_i \Phi_i + \int_\Gamma \Phi \left(\vec{n} \cdot \nabla G_{i,j} \right) d\Gamma = \int_\Gamma G_{i,j} \left(\vec{n} \cdot \nabla \Phi \right) d\Gamma. \tag{25}$$

For a smooth surface at the point i, the constant c_i is equal to π [43].

3.1.1 Infinite domains
Everything done to this point has been done under the assumption that the boundary encloses the domain. In many situations, the domain lies outside the boundary. In order to get a solution for this situation, it is necessary to introduce a second boundary, $\bar{\Gamma}$, placed around the surface of interest and centered on the source point on that surface. Adding the enclosing boundary to eqn (25), one obtains

$$\begin{aligned} c_i \Phi_i + \int_\Gamma \Phi \left(\vec{n} \cdot \nabla G_{i,j} \right) d\Gamma + \int_{\bar{\Gamma}} \Phi \left(\vec{n} \cdot \nabla G_{i,j} \right) d\bar{\Gamma} \\ = \int_\Gamma G_{i,j} \left(\vec{n} \cdot \nabla \Phi \right) d\Gamma + \int_{\bar{\Gamma}} G_{i,j} \left(\vec{n} \cdot \nabla \Phi \right) d\bar{\Gamma}. \end{aligned} \tag{26}$$

If the radius of the new boundary is taken to infinity then the limit of the integrals over the external boundary, $\bar{\Gamma}$, go to infinity

$$\lim_{R \to \infty} \left(\int_{\bar{\Gamma}} \Phi \left(\vec{n} \cdot \nabla G_{i,j} \right) d\bar{\Gamma} - \int_{\bar{\Gamma}} G_{i,j} \left(\vec{n} \cdot \nabla \Phi \right) d\bar{\Gamma} \right) = H_{i,\infty}. \tag{27}$$

The value of eqn (27) can be found analytically

$$- \int_{\Gamma_\infty} \Phi \left(\vec{n} \cdot \nabla G_{i,j} \right) d\Gamma_\infty = H_{i,\infty} = 4\pi \Phi \tag{28}$$

where the minus sign indicates the direction of the normal vector. The second term in eqn (27) vanishes as the limit is taken. The value of integral in (28) can be more easily seen if it is transformed to spherical coordinates

$$\lim_{r \to \infty} \left(- \int_0^\pi \int_0^{2\pi} \frac{\Phi}{r^2} \left(r^2 \sin \phi \right) d\theta d\phi \right) = 4\pi \Phi \tag{29}$$

where r is also the normal vector since the surface is a sphere centered at a source point. Φ at infinity is often assumed to be zero; thus satisfying the zero radiation condition at infinity exactly.

3.1.2 Half spaces
A half space is simply an infinite domain split by a plane. The half space is the space lying on one side of the plane. If either the Dirichlet or Neumann condition vanishes at the plane, the Green's function presents an interesting opportunity to satisfy that condition exactly with a very small additional computation when evaluating the

kernels of the integrals. This is done by making use of the reflection properties of Green's functions [43–46]. In the case of buried pipelines, the Neumann condition vanishes at the plane boundary, *i.e.* there is no current flowing out of the soil into the air, and none is flowing from the air into the soil.

If one starts with the boundary condition on the plane being zero normal current and places a source at x and its reflection about the plane at x' [46],

$$\sigma\left(x\right)\left(\vec{n}\cdot\nabla G\left(x,\xi\right)\right)+\sigma\left(x'\right)\left(\vec{n}\cdot\nabla G\left(x',\xi\right)\right)=0 \tag{30}$$

which implies that the two source intensities $\sigma(x)$, and $\sigma(x')$ are equal and have the same sign. The sign is the same because the outward normal vectors have opposite signs for the z component for the reflections. The final form of the Green's function is

$$G_{i,j}=\frac{1}{4\pi r\left(x_i,x_j\right)}+\frac{1}{4\pi r\left(x_i,x_j'\right)} \tag{31}$$

where x_j' is the reflected source point.

3.1.3 Layers
Layers are created using the same types of Green's function reflections as described above. The only restriction is that one of the two boundary conditions at the interface between the two layers must be equal to zero (*i.e.* either $\Phi=0$ or $\vec{n}\cdot\nabla\Phi=0$) [46]. In the context of cathodic protection of buried pipelines, the only boundary condition that has physical meaning is a zero normal current condition since there is no easy way to have an arbitrary plane within the electrolyte that has a potential equal to zero. Including a plane with a zero Neumann (natural) condition implies that there is one region in which current may flow that is bounded by regions of zero conductivity whose boundaries are defined by the Green's function reflections. An example would be an underlying rock layer which has zero conductivity.

The resulting functions can be obtained by using equations of the form of eqn (30). The Green's function in three dimensions would then be of the form

$$G_{i,j}=\sum_{k}^{\text{Reflections}+1}\frac{1}{4\pi r\left(x_i,x_{j,k}\right)} \tag{32}$$

where the index $k>1$ refers to the reflection about some plane k. For $k=1$, $x_{j,k}$ is the field point on the real object. If the soil surface is the only reflection used, the result is the same as eqn (31).

An example of a pipeline in a halfspace with an underlying rock layer is shown in fig. 2. Two reflections were used. The first accounts for the zero normal current at the air soil interface which is represented by the pipe in the air. The second reflection accounts for the zero normal current at the rock soil interface and is represented by the lower pipe.

The influence of the nonconducting layer on the current delivered to a pipe is seen in fig. 3 for a sequence of calculations for which the distance r between the horizontal pipe and a non-conducting layer placed below the pipe was varied.

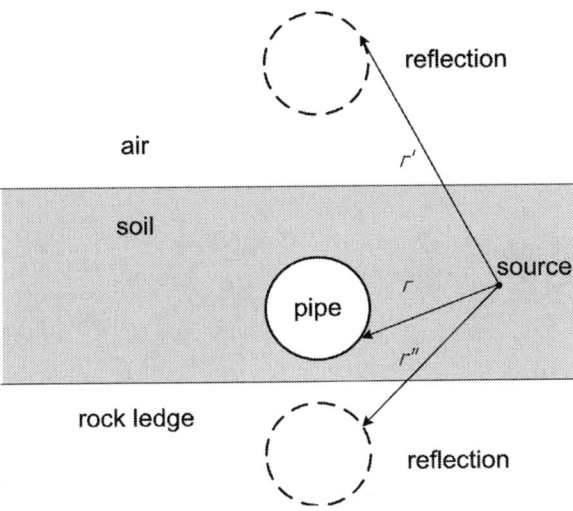

Figure 2: Reflections of Green's function to account for boundaries with zero normal current.

For these calculations, the anode was placed far from the pipe such that the current distribution around the circumference of the pipe was affected only by the screening of the insulating ledge. The current increased with the distance according to

$$\frac{I}{I_{r\to\infty}} = 1 - \frac{a}{\sqrt{r}} \tag{33}$$

where a is a fitted parameter that depends on the geometry and soil resistance and $I_{r\to\infty}$ is the value obtained from Dwight's formula [3, 47] which does not account for the presence of a nonconducting layer. The total current tended toward that predicted by Dwight's formula as the distance between the ledge and the pipe increased.

3.2 Inner domain: finite element method

The Finite Element Method was used for the domain consisting of the pipe-line steel, copper connection wires, and anode material. The Finite Element Method is ideal for completely bounded domains where the material properties may change from one location to another, *i.e.* current flow from the pipe to the copper wire connecting the pipe to the anode. A significant advantage of the formulation was that the same discretization could be employed as was used for the Boundary Element Method solution of the outer domain.

The development of the finite element method for the pipe steel domain starts by writing a weighted residual for the strong form of Laplace's equation given in eqn (7)

$$\int_{\Omega} w \nabla \cdot \vec{\kappa} \cdot \nabla V \, d\Omega = 0 \tag{34}$$

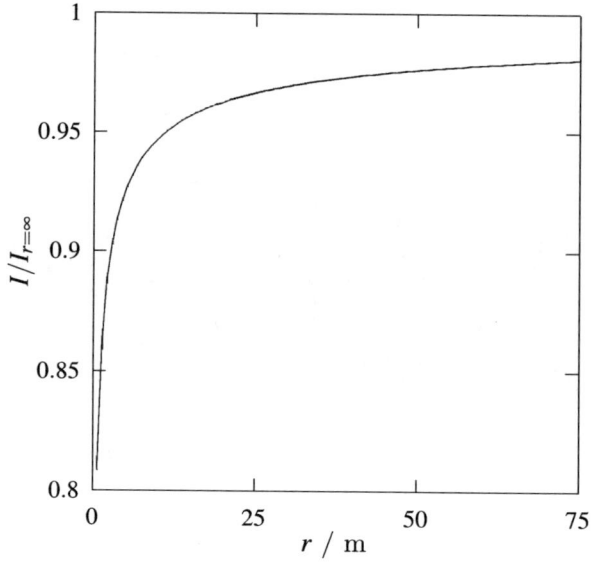

Figure 3: Total calculated current, normalized by the result from Dwight's formula for an infinite soil domain [3,47], as a function of the separation r between the pipe and an underlying (insulating) rock layer.

where w is any weighting function. For pipe steel, κ is just a constant times an identity matrix

$$\vec{\vec{\kappa}} = \kappa \begin{bmatrix} 1 & 0 & 0 \\ 0 & 1 & 0 \\ 0 & 0 & 1 \end{bmatrix} \tag{35}$$

with step changes in κ when the type of metal changes. After substituting the material properties into eqn (34) and integrating by parts using the divergence theorem, the following Weak Form is obtained

$$\iiint_{\Omega} \kappa \nabla w \cdot \nabla V \, dx \, dy \, dz = -\iint w \kappa \, (\vec{n} \cdot \nabla V) \, ds \tag{36}$$

where $\vec{n}$ is the outward normal vector from the boundary of the domain. The domain is divided into elements using piecewise continuous polynomial isoparametric shape functions. The shape functions approximate the geometry and solution, V, over the elements. The discretized form of eqn (36) is

$$\sum_{j=1}^{n} \left[\iiint_{(e)} \kappa \nabla \phi_i^{(e)} \cdot \nabla \phi_j^{(e)} \, dx \, dy \, dz \right] V_j = -\iint \kappa \phi_i^{(e)} \, (\vec{n} \cdot \nabla V)_i^{(e)} \, ds \tag{37}$$

where κ is the scalar component of the property tensor, ϕ_i is a shape function and the sum goes from 1 to the total of all the shape functions of all the elements.

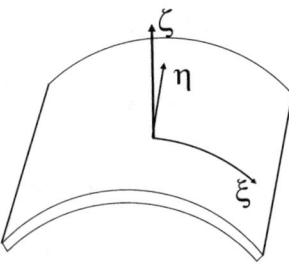

Figure 4: Diagram of the shell element used to calculate the potential drop within the pipe steel. The element is assumed to have no variation in the ζ direction. The curvilinear coordinate system is displayed on top of the element.

3.2.1 Pipe shell elements

A special type of thin shell elements is introduced here and shown in fig. 4. These elements are specifically designed for potential problems on shells where the absolute value of the material property (κ in eqn (37)) is large.

The elements are defined in orthonormal curvilinear coordinates, ξ, η, and ζ where ξ and η define the outside surface of the pipe and ζ is the outwardly directed normal vector. ζ is obtained through the cross product $\xi \times \eta$.

Variations of the potential in the ζ direction were assumed to be negligible because the scale of the problem is many orders of magnitude greater in the ξ and η directions. Variations of the potential parallel to the surface were allowed to vary in a piece-wise continuous way using bi-quadratic shape functions for the elements. These functions were obtained through the product of two Lagrange interpolating polynomials of the same order, one in ξ and one in η. The result is a family of elements with square parent elements which include the four-node linear, the nine-node quadratic, and the 16 node cubic elements.

3.2.2 Applying elements to the FEM

The integral (37) must be transformed to the curvilinear coordinate system of the parent elements. A differential volume, $dx\,dy\,dz$ can be transformed to the curvilinear system by the determinate of the Jacobian of the coordinate transformation

$$dx\,dy\,dz = |\mathbf{J}|\,d\xi\,d\eta\,d\zeta \tag{38}$$

where $\mathbf{J}$ is the Jacobian.

The Jacobian is composed of the partial derivatives of the coordinates x, y and z with respect to each of the curvilinear coordinates

$$\mathbf{J} = \begin{bmatrix} \frac{\partial x}{\partial \xi} & \frac{\partial y}{\partial \xi} & \frac{\partial z}{\partial \xi} \\ \frac{\partial x}{\partial \eta} & \frac{\partial y}{\partial \eta} & \frac{\partial z}{\partial \eta} \\ \frac{\partial x}{\partial \zeta} & \frac{\partial y}{\partial \zeta} & \frac{\partial z}{\partial \zeta} \end{bmatrix}. \tag{39}$$

Integral (37) is transformed to the curvilinear coordinate system to yield

$$\sum_i \left[\int\int\int_{\Omega^{(e)}} \sum_k \frac{\partial \phi_i}{\partial x_k} \kappa \cdot \frac{\partial \phi_j}{\partial x_k} |\mathbf{J}| d\xi \, d\eta \, d\zeta \right] = - \oiint_{\Gamma_{(e)}} w\kappa \, (\vec{n} \cdot \nabla V) \, ds \quad (40)$$

where x_k is one of the cartesian coordinates, $|\mathbf{J}|$ is the determinate of the Jacobian, s is the surface of the element $\vec{n}$ is the outward normal vector from the surface, and $\Omega^{(e)}$ is the domain of the parent element. The limits of integration in the parent element are from -1 to $+1$ for all three of the coordinates. For the special shell elements used for pipes, the integral is only performed numerically over the surface $d\xi \, d\eta$ which is then scaled by the physical thickness of the shell h, the result of the integral over ζ. This requires the Jacobian to be modified to a surface Jacobian which, in this case, is simply the square root of the magnitude of the Jacobian. Rewriting eqn (40), the two-dimensional integral over the surfaces of the pipes is obtained as

$$\sum_i \left[h \int\int_{\Omega^{(e)}} \sum_k \frac{\partial \phi_i}{\partial x_k} \kappa \cdot \frac{\partial \phi_j}{\partial x_k} \sqrt{|\mathbf{J}|} d\xi \, d\eta \right] = - \oiint_{\Gamma_{(e)}} w\kappa \, (\vec{n} \cdot \nabla V) \, ds \quad (41)$$

with the thickness of the steel, h, a parameter. The normal vector has the same direction as the one from the boundary element method.

Partial derivatives in eqn (41) with respect to cartesian coordinates need to be expressed in terms of the curvilinear coordinates. Using the chain rule, but starting from the derivatives in cartesian coordinates, one writes

$$\left\{ \begin{array}{c} \frac{\partial \phi}{\partial x} \frac{\partial x}{\partial \xi} + \frac{\partial \phi}{\partial y} \frac{\partial y}{\partial \xi} + \frac{\partial \phi}{\partial z} \frac{\partial z}{\partial \xi} \\ \frac{\partial \phi}{\partial x} \frac{\partial x}{\partial \eta} + \frac{\partial \phi}{\partial y} \frac{\partial y}{\partial \eta} + \frac{\partial \phi}{\partial z} \frac{\partial z}{\partial \eta} \\ \frac{\partial \phi}{\partial x} \frac{\partial x}{\partial \zeta} + \frac{\partial \phi}{\partial y} \frac{\partial y}{\partial \zeta} + \frac{\partial \phi}{\partial z} \frac{\partial z}{\partial \zeta} \end{array} \right\} = \left\{ \begin{array}{c} \frac{\partial \phi}{\partial \xi} \\ \frac{\partial \phi}{\partial \eta} \\ \frac{\partial \phi}{\partial \zeta} \end{array} \right\} \quad (42)$$

or, in terms of the Jacobian

$$\mathbf{J} \left\{ \begin{array}{c} \frac{\partial \phi}{\partial x} \\ \frac{\partial \phi}{\partial y} \\ \frac{\partial \phi}{\partial z} \end{array} \right\} = \left\{ \begin{array}{c} \frac{\partial \phi}{\partial \xi} \\ \frac{\partial \phi}{\partial \eta} \\ \frac{\partial \phi}{\partial \zeta} \end{array} \right\}. \quad (43)$$

Since ϕ is a function of ξ, η, and ζ, the partial derivatives of ϕ with respect to the cartesian coordinates can be found by inverting the Jacobian.

$$\left\{ \begin{array}{c} \frac{\partial \phi}{\partial x} \\ \frac{\partial \phi}{\partial y} \\ \frac{\partial \phi}{\partial z} \end{array} \right\} = \mathbf{J}^{-1} \left\{ \begin{array}{c} \frac{\partial \phi}{\partial \xi} \\ \frac{\partial \phi}{\partial \eta} \\ \frac{\partial \phi}{\partial \zeta} \end{array} \right\}. \quad (44)$$

If $\mathbf{J}_{i,j}^{-1}$ is the element from the ith row and the jth column of the inverse of the Jacobian, then the partial derivatives defined in eqn (44) can be expanded to yield

$$\frac{\partial \phi_i}{\partial x} = (J^{-1})_{11} \frac{\partial \phi_i}{\partial \xi} + (J^{-1})_{12} \frac{\partial \phi_i}{\partial \eta} + (J^{-1})_{13} \frac{\partial \phi_i}{\partial \zeta}$$

$$\frac{\partial \phi_i}{\partial y} = (J^{-1})_{21} \frac{\partial \phi_i}{\partial \xi} + (J^{-1})_{22} \frac{\partial \phi_i}{\partial \eta} + (J^{-1})_{23} \frac{\partial \phi_i}{\partial \zeta} \qquad (45)$$

$$\frac{\partial \phi_i}{\partial z} = (J^{-1})_{31} \frac{\partial \phi_i}{\partial \xi} + (J^{-1})_{32} \frac{\partial \phi_i}{\partial \eta} + (J^{-1})_{33} \frac{\partial \phi_i}{\partial \zeta}.$$

The assumption that the thickness of the shell is small with respect to the radius and length of the shell means that there is negligible variation in the potential in the ζ direction. Therefore, all terms in eqn (45) that involve partial derivatives with respect to ζ can be assumed to be equal to zero. All of the terms in eqn (41) can be evaluated numerically. Since the shape functions for the geometry and solution vary quadratically in the ξ and η directions, a nine-point Gauss rule (3×3) applied in both directions will give an exact (to machine precision) result.

3.2.3 Bonds and resistors

Bonds and resistors are used to provide paths for flow of electrical current between pipes or between pipes and anodes. Bonds were implemented as a linear 1D element between the connection node on one pipe or anode to the connection node on the second pipe or anode. This formulation requires introduction of no additional nodes. The material property is set by accounting for the real length and gauge of the wire that ties the pipes together. If a resistor is specified within the wire, it is added to the total resistance of the bond. An illustration of a bond is given in fig. 5. The lines connecting the pipes to each other and to the anode are bonds. The cylinder on the line connecting the two pipes represents a resistor that is sometimes added in series with the bond to adjust the current load applied to the pipelines. They are modeled as 1D finite elements that connect to the bond connection points. The value of the resistor is set to zero if a resistor is not used in the calculation.

A one-dimensional version of Laplace's equation is used for the 1D elements. The development for the linear 1D element follows that presented in section 3.2. The result of the integration over the bond is the total conductance of the bond or

$$K_e = \begin{bmatrix} \frac{1}{R} & \frac{-1}{R} \\ \frac{-1}{R} & \frac{1}{R} \end{bmatrix} \qquad (46)$$

where R is the total resistance of the bond.

3.3 Coupling BEM to FEM

The finite-element model for the inner domain is coupled to the boundary-element for the outer domain at the interface between the two domains. Ohm's Law holds within each domain as stated in section 2.1. Therefore, at any arbitrary surface that forms a boundary between two domains it can be shown that the flux on either side

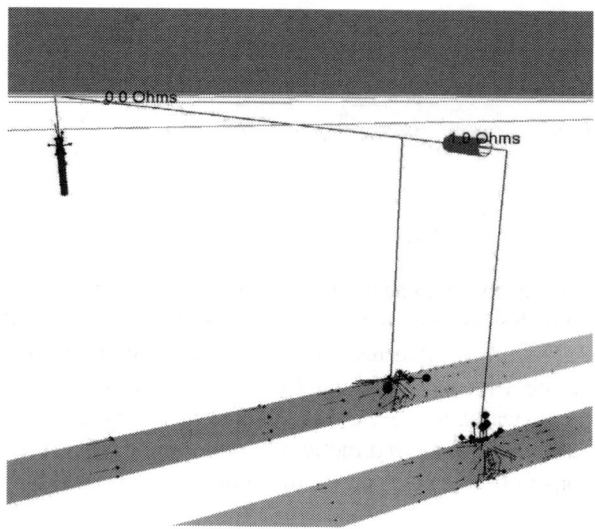

Figure 5: A representation of two horizontal pipes connected to each other and to a vertical anode or ground bed. The lines connecting the pipes to each other and to the anode are bonds. The cylinder on the line connecting the two pipes represents a resistor. They are modeled as 1D finite elements that connect to the bond connection points.

of the boundary is related by the material property,

$$\kappa_1 \vec{n} \cdot \nabla \Phi_1 = \kappa_2 \vec{n} \cdot \nabla \Phi_2 \tag{47}$$

which is a balance on charge at the pipe/soil interface, where κ is the material property. For potential problems, κ is the conductivity of the material in Mho/m. Using the variables for potential in the pipe, V, and potential in the soil, Φ, the interface condition is written as

$$(\vec{n} \cdot \nabla V) = \frac{\kappa_{\text{soil}}}{\kappa_{\text{steel}}} (\vec{n} \cdot \nabla \Phi) \tag{48}$$

where the quantity $\vec{n} \cdot \nabla V$ is used in the finite element load integral in eqn (41). Equation (48) can be inserted into eqn (41) to obtain

$$\sum_i \left[h \int\!\!\int_{\Omega^{(e)}} \sum_k \frac{\partial \phi_i}{\partial x_k} \kappa \cdot \frac{\partial \phi_j}{\partial x_k} \sqrt{|\mathbf{J}|} d\xi \, d\eta \right] = - \int\!\!\int_{\Gamma_{(e)}} w \kappa_{\text{soil}} (\vec{n} \cdot \nabla \Phi) \, ds. \tag{49}$$

The right side of eqn (49) links the soil domain to the steel domain through the current density generated by the kinetics of the corrosion reaction at the steel surface given by eqns (9) and (12).

3.4 Discretization of the boundary elements

Equation (25) may now be applied to a surface that has its boundaries broken up into individual finite elements. The solution to the differential equation can be represented by approximate functions on each individual element. Example functions may be 0th order or constant value, 1st order or linear, *etc.* The simplest case is the constant element. The value of Φ is assumed to be constant across each element. Then one can write an equation of the form of (25) for each degree of freedom that makes up the surface and where the subscript i refers to the element number. The integrals are broken up into N sub-intervals corresponding to N elements that when summed together form the complete integral over the entire boundary

$$c_i \phi_i + \sum_j \int_{\Gamma_j} \Phi \left(\vec{n} \cdot \nabla G_{i,j} \right) d\Gamma_j = \sum_j \int_{\Gamma_j} G_{i,j} \left(\vec{n} \cdot \nabla \Phi \right) d\Gamma_j. \tag{50}$$

When constant element are used, the unknown element potentials and normal current densities can be brought outside the integrals. Since a well-posed problem has one half of the boundary conditions specified, the result is N equations with N unknowns.

Since it is desired to be able to change the boundary condition type as well as value, the notation used here will denote the matrix resulting from the left hand integral as H and the right side as G

$$H\Phi = G \left(\vec{n} \cdot \nabla \Phi \right) \tag{51}$$

where $\vec{n} \cdot \nabla \Phi$ is the normal current density divided by the conductivity at the boundary.

3.5 Self-equilibration

Cathodic protection systems are constrained by the requirement that charge is conserved. The boundary element formulation developed in section 2.1 requires modification to satisfy the constraint that charge is conserved. Instead of having a specified potential at infinity (most often set to zero) which serves as a source or sink for charge, an unknown value of potential at infinity is used such that no current enters or leaves through that boundary. To implement this condition, an extra equation is added to the system [36, 48]

$$\sum_i \int_{\Gamma_i} \kappa \hat{\vec{n}} \cdot \nabla \Phi d\Gamma_i = 0. \tag{52}$$

Equation (52) simply states that all currents entering the boundaries of the domain sum to zero, or, in other words, no current is lost to or gained from infinity. The left hand side is added as a new row at the bottom of the G matrix of eqn (51), while the matrix H in (51) receives a row of zeros. A column is added to the H matrix which corresponds to the unknown potential at infinity. The values that are placed in this column come from eqn (28) and are all 4π or 1 depending on where the 4π from the Green's function is placed. This would make the H matrix singular because there is still a row of zeros in it. However, that would only be the case if Neumann type boundary conditions are specified everywhere. The Neumann problem results in an infinite number of solutions that differ by a constant. Therefore, at least one element in the system must have a Dirichlet boundary condition to make the H matrix nonsingular and result in a unique solution.

3.6 Multiple CP systems

Riemer and Orazem describe in detail the development needed to model interactions among CP systems [25]. To allow stray current between separate CP systems, additional rows can be added of the type introduced in eqn (52) [49]. For each separate CP system, one extra column in the H matrix is also added. The appearance of the matrices for 2 CP systems in the same domain will be

$$G = \begin{bmatrix} G_{1,1} & G_{1,2} & G_{1,3} & G_{1,4} \\ G_{2,1} & G_{2,2} & G_{2,3} & G_{2,4} \\ 0 & 0 & A_3 & A_4 \\ A_1 & A_2 & 0 & 0 \end{bmatrix} \tag{53}$$

$$H = \begin{bmatrix} H_{1,1} & H_{1,2} & 0 & 1 \\ H_{2,1} & H_{2,2} & 1 & 0 \\ 0 & 0 & 0 & 0 \\ 0 & 0 & 0 & 0 \end{bmatrix} \tag{54}$$

with the column matrix u given by

$$u = \begin{bmatrix} u_1 \\ u_2 \\ u_{\infty,\text{System2}} \\ u_{\infty,\text{System1}} \end{bmatrix} \tag{55}$$

and the column matrix q defined in the usual way. When there are two or more CP systems within a domain, a Dirichlet condition on at least one element in each CP system must be specified to prevent the H matrix from being singular. The added equations and unknown potentials at infinity are sufficient to allow several CP systems to interact with each other while enforcing that the total current on each CP system sum to zero.

3.7 Nonlinear boundary conditions with attenuation in the pipe steel

The external load integral from the finite-element solution to the inner domain can be written for each node as a sum of contributions from each element that has that node in common

$$f_i = \sum_{\ell=1}^{nce} - \iint_{\Gamma_{(e)}} w_{\ell,i} \kappa_{\text{soil}} \left(\vec{n} \cdot \nabla \Phi \right) ds \tag{56}$$

where f_i is the load for node i and nce is the number of contributing elements to the load at node i. This integral must be evaluated across all nodes on all surfaces of all structures in the model. Since the value of $\vec{n} \cdot \nabla \Phi$ is either a known or unknown boundary condition from the boundary element method, it must be factored out of the integral. Using the shape functions that describe the values of $\vec{n} \cdot \nabla \Phi$ across each element in terms of the nodal values $\vec{n} \cdot \nabla \Phi$ and the weights, one can rewrite eqn (56) as

$$f_i = \sum_{\ell=1}^{nce} - \iint_{\Gamma_{(e)}} w_{\ell,i} \kappa_{\text{soil}} \sum_{k=1}^{nen} \phi_{\ell,k}(s) \left(\vec{n} \cdot \nabla \Phi \right)_{\ell,k} ds \tag{57}$$

where nen is the number of nodes in the element, $(\vec{n} \cdot \nabla \Phi)_k$ is the nodal value of the normal electric field within the element n, and $\phi(s)_k$ is the shape function whose value is one at node k. Using the property of integrals that an integral of a sum is the sum of integrals, eqn (57) can be rewritten as

$$f_i = \sum_{\ell=1}^{nce} \sum_{k=1}^{nen} - \iint_{\Gamma_{(e)}} w_{\ell,i} \kappa_{\text{soil}} \phi_{\ell,k}(s) \left(\vec{n} \cdot \nabla \Phi \right)_{\ell,k} ds. \tag{58}$$

Equation (58) can then be written in matrix form as

$$f_i = \left[\ldots - \iint_{\Gamma_{(e)}} w_{\ell,i} \kappa_{\text{soil}} \phi_{\ell,k}(s) ds \ldots \right] \begin{bmatrix} (\vec{n} \cdot \nabla \Phi)_{1,1} \\ \vdots \\ (\vec{n} \cdot \nabla \Phi)_{ne,nen} \end{bmatrix} \tag{59}$$

or

$$f_i = \hat{\boldsymbol{F}}_i \left[\vec{n} \cdot \nabla \Phi \right]. \tag{60}$$

These sub-matrices are then assembled with the right hand column matrix from eqn (51) to form the global finite element system

$$\boldsymbol{K}\left[V\right] = \hat{\boldsymbol{F}}\left[\vec{n} \cdot \nabla \Phi\right] \tag{61}$$

where $\hat{\boldsymbol{F}}$ is a matrix formed from the assembly of eqn (60) such that the column matrix corresponding to the Boundary Element Method is formed on the right hand side. $\hat{\boldsymbol{F}}$ is not symmetric.

4 Method for solution of systems of nonlinear equations

The set of variables that appear in the combined soil domain metal domain problem are potentials outside the surface of all pipes or tanks (Φ), unknown normal electric field on anodes ($\vec{n} \cdot \nabla \Phi$), and unknown potential difference ($V - \Phi$) for coated and bare protected structures. Introduction of the attenuation code from eqn (61) to the problem introduces n more equations and n more unknowns where n is the number of nodes used to describe the boundary mesh for the problem. The total number of algebraic equations to be solved is now $2n + k$, where k is the number of separate CP systems.

4.1 Global matrix

The coefficients obtained from the coupled methods are placed into a global matrix system and one instance of eqn (52) is added for self-equilibriation of each CP system

$$
\begin{bmatrix}
\boldsymbol{H}_{c,c} & \boldsymbol{H}_{a,c} & 0 & 0 & -4\pi \\
\boldsymbol{H}_{c,a} & \boldsymbol{H}_{a,a} & 0 & 0 & -4\pi \\
0 & 0 & \boldsymbol{K}_c & 0 & 0 \\
0 & 0 & 0 & \boldsymbol{K}_a & 0 \\
0 & 0 & 0 & 0 & 0
\end{bmatrix}
\begin{bmatrix}
\Phi_c \\
\Phi_a \\
V_p \\
V_a \\
\Phi_\infty
\end{bmatrix}
$$

$$
=
\begin{bmatrix}
\boldsymbol{G}_{c,c} & \boldsymbol{G}_{a,c} \\
\boldsymbol{G}_{c,a} & \boldsymbol{G}_{a,a} \\
\hat{\boldsymbol{F}}_c & 0 \\
0 & \hat{\boldsymbol{F}}_a \\
A_c & A_a
\end{bmatrix}
\begin{bmatrix}
\vec{n} \cdot \nabla \Phi_c \\
\vec{n} \cdot \nabla \Phi_a
\end{bmatrix}
\tag{62}
$$

where the subscripts a and c refer to anodes and cathodes (pipes), respectively, double subscripts are from the boundary element method and refer to where the source point and field point lie, respectively, and A is the area of each section.

The columns of eqn (62) are sorted such that all of the unknown variables are on the left hand side. For anodes, the unknown variable is the current density ($\vec{n} \cdot \nabla \Phi$), and for all other structures it is the potential (Φ). The reordered system then looks like

$$
\begin{bmatrix}
H_{c,c} & -G_{a,c} & 0 & 0 & -4\pi \\
H_{c,a} & -G_{a,a} & 0 & 0 & -4\pi \\
0 & 0 & K_c & 0 & 0 \\
0 & -\hat{F}_a & 0 & K_a & 0 \\
0 & -A_a & 0 & 0 & 0
\end{bmatrix}
\begin{bmatrix}
\Phi_c \\
\vec{n} \cdot \nabla \Phi_a \\
V_c \\
V_a \\
\Phi_\infty
\end{bmatrix}
$$

$$
=
\begin{bmatrix}
G_{c,c} & -H_{a,c} \\
G_{c,a} & -H_{a,a} \\
\hat{F}_c & 0 \\
0 & 0 \\
A_c & 0
\end{bmatrix}
\begin{bmatrix}
\vec{n} \cdot \nabla \Phi_c \\
\Phi_a
\end{bmatrix}
\tag{63}
$$

where the column matrix on the left hand side is the set of unknown variables and the column matrix on the right side is the set of known boundary conditions. Equation (63) can be rewritten as

$$
[A]
\begin{bmatrix}
\Phi_c \\
\vec{n} \cdot \nabla \Phi_a \\
V_c \\
V_a \\
\Phi_\infty
\end{bmatrix}
= [B]
\begin{bmatrix}
\vec{n} \cdot \nabla \Phi_c \\
\Phi_a
\end{bmatrix}
\tag{64}
$$

where A and B correspond to the matrices in eqn (63).

If the boundary conditions are constant, the unknown variables can be obtained by a simple matrix inversion. For problems in electrochemistry, the set of known boundary conditions are given as a set of nonlinear functions of the unknown variables. Therefore, the solution must be obtained using a technique for coupled sets of nonlinear algebraic equations. This is done by rewriting eqn (64) as a function that is equal to zero

$$
[A]
\begin{bmatrix}
\Phi_c \\
\vec{n} \cdot \nabla \Phi_a \\
V_c \\
V_a \\
\Phi_\infty
\end{bmatrix}
- [B]
\begin{bmatrix}
\vec{n} \cdot \nabla \Phi_c = f(V, \Phi_c) \\
\Phi_a = f(V, \vec{n} \cdot \nabla \Phi_a)
\end{bmatrix}
=
\begin{bmatrix}
0 \\
0 \\
0 \\
0 \\
0
\end{bmatrix}.
\tag{65}
$$

Since one of the column matrices contains unknown variables, a guess for the values must be made. Since the guess will not be the correct solution to eqn (65), the right hand column vector will not be equal to zero, but a residual

$$[A] \begin{bmatrix} \Phi_c \\ \vec{n} \cdot \nabla \Phi_a \\ V_c \\ V_a \\ \Phi_\infty \end{bmatrix} - [B] \begin{bmatrix} f(V, \Phi_c) \\ f(V, \vec{n} \cdot \nabla \Phi_a) \end{bmatrix} = \begin{bmatrix} R \end{bmatrix}. \tag{66}$$

Rewriting eqn (65) as a general set of equations of an unknown vector x

$$F(x) = 0 \tag{67}$$

a Jacobian J, of the set of equations can be written where an element in the Jacobian is given by

$$J_{i,j} = \frac{\partial F_i}{\partial x_j} \bigg|_{x_{\ell \neq j}}. \tag{68}$$

Then the Newton–Raphson iteration can be performed to obtain the solution vector x such that eqn (67) holds. Again the line search is necessary for the the method to work. However, in most non-trivial cases, even the line searches fail.

In order to get the method to converge to a solution, many techniques were tried including the minimal residual techniques such as the modified Powell method [50], Levenberg–Marquort techniques and other methods such as the conjugate gradient method. These techniques would converge sometimes if the guess was nearly the correct value. However, in general, the convergence behavior was very poor.

4.2 Variable transformation to stabilize convergence

The system of equation that result from Φ, V, i as the set of variables for the solution has very poor convergence characteristics. Great effort was required to bring the system to convergence. Standard line searches and even minimal residual techniques failed to converge.

A simple variable transformation was developed that enables good convergence properties using the simple line search method. The variable transformation needed can be readily identified from the equations such as (9) used to describe the reaction kinetics. In eqn (9), the difference $V - \Phi$ appears many times. A simple transformation can be written

$$\Psi = V - \Phi \tag{69}$$

where Ψ represents the driving force for the electrochemical kinetics. The variable V is then chosen as the dependent variable and rewritten as

$$V = \Psi + \Phi. \tag{70}$$

All of the equations used for the kinetics, *i.e.* eqn (9) for bare metal, eqn (12) for coated metal, eqn (13) for galvanic anodes, and eqn (16) for impressed current anodes, can be rewritten in terms of the new variable. For bare metal, one obtains

$$i = 10^{(\Psi - E_{Fe})/\beta_{Fe}} - \left(\frac{1}{i_{lim,O_2}} + 10^{(\Psi - E_{O_2})/\beta_{O_2}} \right)^{-1} - 10^{-(\Psi - E_{H_2})/\beta_{H_2}} \quad (71)$$

For coated metal, two simultaneous equations in two unknowns describe the current density, *i.e.*

$$i = \frac{-(\Phi - \Phi_{in})}{\rho_{film} \delta_{film}} \quad (72)$$

and

$$i = \frac{A_{pore}}{A} \left[10^{(\Psi + \Phi - \Phi_{in} - E_{Fe})/\beta_{Fe}} \right.$$
$$- \left(\frac{1}{(1 - \alpha_{blk})i_{lim,O_2}} - 10^{(\Psi + \Phi - \Phi_{in} - E_{O_2})/\beta_{O_2}} \right)^{-1}$$
$$\left. - 10^{-(\Psi + \Phi - \Phi_{in} - E_{H_2})/\beta_{H_2}} \right]. \quad (73)$$

For galvanic anodes,

$$\Psi = E_{anode} + \beta_{anode} \log(i/i_{corr} + 1) \quad (74)$$

and for impressed current anodes,

$$\Psi = \triangle V_{rectifier} + E_{O_2} + \beta_{O_2} \log(i/i_{corr} + 1) \quad (75)$$

where $\triangle V_{rectifier}$ is the known setting for the rectifier.
   The set of equations for the finite element method must also be rewritten as

$$K[\Psi + \Phi] = [f] \quad (76)$$

where f and K do not change. The boundary element method is not changed. The set of equations that is solved becomes:

$$
\begin{bmatrix}
0 & 0 & \boldsymbol{H}_{c,c} & \boldsymbol{H}_{a,c} & -4\pi \\
0 & 0 & \boldsymbol{H}_{c,a} & \boldsymbol{H}_{a,a} & -4\pi \\
\boldsymbol{K}_c & 0 & \boldsymbol{K}_c & 0 & 0 \\
0 & \boldsymbol{K}_a & 0 & \boldsymbol{K}_a & 0 \\
0 & 0 & 0 & 0 & 0
\end{bmatrix}
\begin{bmatrix}
\Psi_c \\
\Psi_a \\
\Phi_c \\
\Phi_a \\
\Phi_\infty
\end{bmatrix}
$$

$$
=
\begin{bmatrix}
\boldsymbol{G}_{c,c} & \boldsymbol{G}_{a,c} \\
\boldsymbol{G}_{c,a} & \boldsymbol{G}_{a,a} \\
\hat{\boldsymbol{F}}_c & 0 \\
0 & \hat{\boldsymbol{F}}_a \\
\boldsymbol{A}_c & \boldsymbol{A}_a
\end{bmatrix}
\begin{bmatrix}
\vec{n} \cdot \nabla \Phi_c \\
\vec{n} \cdot \nabla \Phi_a
\end{bmatrix}.
\tag{77}
$$

At this point the essential boundary condition for V ($\Psi + \Phi$) is applied. There is one degree of freedom in the problem for this potential so one node on one pipe is selected to have a 0 potential. This is done by replacing the FEM equation for the single selected node in the system given by eqn (77) with

$$
\Psi + \Phi = 0.
\tag{78}
$$

The FEM equation (lower half of the system) is replaced because that is where the degree of freedom in the reference potential is. If there are multiple CP systems, then there must be an equal number of essential boundary conditions on $\Psi + \Phi$. Once the equations are set up, all unknown variables are placed on the left-hand-side, which are Φ and Ψ for cathodes (pipes, tank bottoms, ship hulls, *etc.*) and $\vec{n} \cdot \nabla \Phi$ and Φ for anodes. The reordered system of equations is

$$
\begin{bmatrix}
0 & -\boldsymbol{G}_{a,c} & \boldsymbol{H}_{c,c} & \boldsymbol{H}_{a,c} & -4\pi \\
0 & -\boldsymbol{G}_{a,a} & \boldsymbol{H}_{c,a} & \boldsymbol{H}_{a,a} & -4\pi \\
\boldsymbol{K}_c & 0 & \boldsymbol{K}_c & 0 & 0 \\
0 & -\hat{\boldsymbol{F}}_a & 0 & \boldsymbol{K}_a & 0 \\
0 & -\boldsymbol{A}_a & 0 & 0 & 0
\end{bmatrix}
\begin{bmatrix}
\Psi_c \\
\vec{n} \cdot \nabla \Phi_a \\
\Phi_c \\
\Phi_a \\
\Phi_\infty
\end{bmatrix}
$$

$$
=
\begin{bmatrix}
\boldsymbol{G}_{c,p} & 0 \\
\boldsymbol{G}_{c,a} & 0 \\
\hat{\boldsymbol{F}}_c & 0 \\
0 & -\boldsymbol{K}_a \\
\boldsymbol{A}_c & 0
\end{bmatrix}
\begin{bmatrix}
\vec{n} \cdot \nabla \Phi_c \\
\Psi_a
\end{bmatrix}.
\tag{79}
$$

This system has excellent convergence characteristics and uses about the same number of iterations as a system without the effects of the pipe metal included.

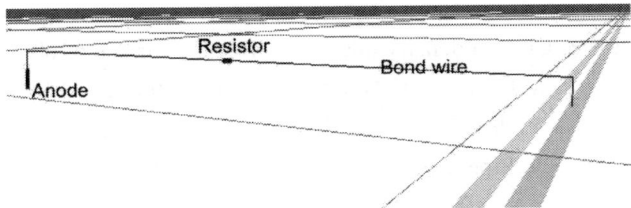

Figure 6: Portion of the configuration of 2 pipes plus 1 of the anodes providing protective current. The display includes an optional resistor in the connecting bond.

5 Examples

The above development provides a foundation for modeling cathodic protection of long stretches of multiple pipelines, including interaction among cathodic protection networks, while retaining the flexibility to account for the role of discrete coating holidays. Care must be taken in the implementation of the approach presented above. Adaptive integration techniques are needed to generate values of sufficient accuracy for the terms appearing in the coefficient matrices. An efficient nonuniform meshing algorithm is needed to avoid numerical errors associated with abrupt changes in mesh size while minimizing the computational cost of the program.

Two examples are included here to show the potential of the model described above. The first shows the calculation of multiple CP systems for two closely spaced pipes in a right-of-way with coating holidays. The second shows how the electrical midpoint may be found.

5.1 Multiple CP Systems

The following example concerns 2 pipelines in a right-of-way protected by separate cathodic protection systems. Three miles of pipe are modeled with both pipes exhibiting several coating holidays. A portion of the model is shown in fig. 6. The smaller pipe is assumed to be older and to have a poor quality coating, while the larger was assumed to have the better coating. No bonds connect the two CP systems.

When the complete system is modeled using the methods given in section 3.6, the current density integrated across the surface of each object (anode or pipe) sum to zero for each CP system, as demonstrated in table 3. As indicated in table 3, Pipe 1, Anode 1, and Anode 3 comprise CP System 1; and Pipe 2 and Anode 2 comprise CP System 2. The amount of current drawn from Anodes 1 and 3 depends on the potential applied and on the quality of the coating on adjacent portions of the pipe.

Even though the two pipelines were electrically isolated from each other, they still influenced the respective level of cathodic protection through the influence of potential gradients in the soil. A plot of current density along the length of one of the two pipes is shown in fig. 7. It is interesting to note that the pipe presented in fig. 7 does not have a holiday at position 1.61km, but the adjacent pipe does. A sharp

Table 3: Current density integrated over the surface of objects.

Object	CP System 1	CP System 2
Pipe 1	−0.40180 A	0 A
Pipe 2	0 A	−0.26287 A
Anode 1	0.35655 A	0 A
Anode 2	0 A	0.26287 A
Anode 3	0.04525 A	0 A
Total	0 A	0 A

decrease in the cathodic current (seen as a spike due to the scale used) appears on the pipe due to the large current demand associated with the coating defect on the adjacent pipe. The same area is shown in a false color plot in fig. 8. A shadow effect is seen on both objects due to the larger current density drawn to the coating defect as compared to that required by the coated regions.

The solution to the potential distribution within the pipe steel allows investigation of the flow of current back to the anodes. The current vectors in the pipe steel in the vicinity of a large coating holiday are presented in fig. 9. The dark rectangle represents a large coating defect which exposes bare steel. The surrounding area is covered by intact coating. The vectors drawn show how the current entering the pipe metal through the holiday spreads within the pipe steel.

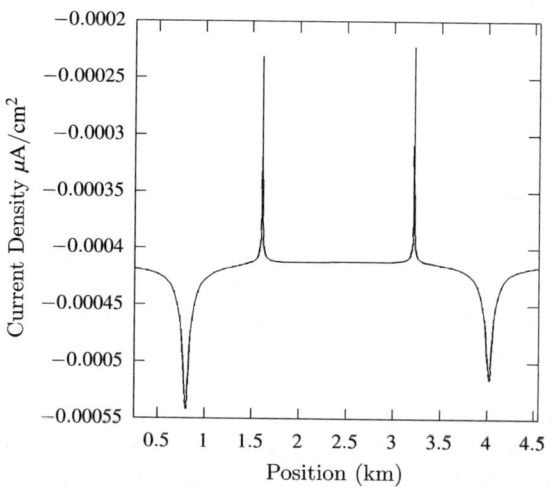

Figure 7: Calculated current density as a function of axial position for Pipe 1. The current density was obtained for the section of pipe directly facing the other pipe.

Figure 8: False color image of current density. The lighter color indicates lower current density.

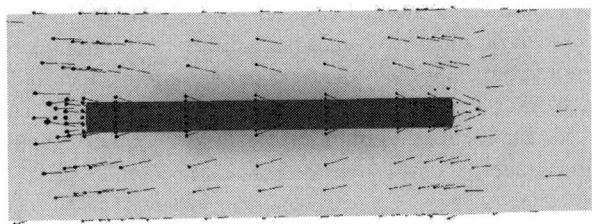

Figure 9: False color image of current flow through the pipe metal. The flow vectors indicate that the current entering the pipe metal through the holiday spreads within the pipe steel.

5.2 Electrical midpoint

Electrical midpoints can also be determined from the steel voltage solution. To demonstrate, a 4-mile long pipeline is modeled with three anodes placed along its

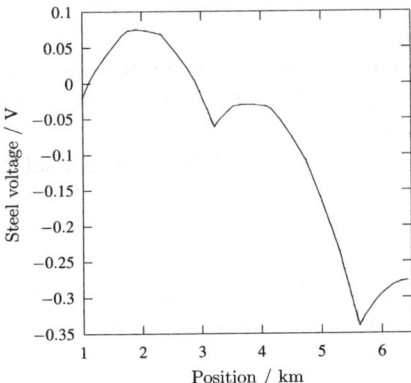

Figure 10: Potential in the steel as a function of axial position. The electrical midpoint can be identified by the maxima between cusps associated with connections to the anodes.

length. The anodes are set to different driving potentials. The potential in the steel, referenced to the potential at the start of the pipe, is presented in fig. 10.

The reference location is given by the node for which eqn (78) applies. The sharp minima or cusps shown in fig. 10 are associated with connections to anodes, ground beds, or other pipes. The electrical midpoints can be identified by the maxima between cusps. As seen in fig. 10, there are two maxima along the length which indicate the locations of the electrical midpoints.

6 Conclusions

This chapter shows how a rigorous first-principles-based model for cathodic protection of multiple pipelines in a right-of-way with coating holidays can be developed which accounts for all effects associated with the interaction of the pipes, anodes and other parts of the CP system. Large shadow effects are seen when coating holidays are included or when two pipes have different coating qualities.

It should be noted the cathodic protection model requires locally valid polarization curves which describe the chemistry at the pipe–soil interface. The polarization curves account for the local soil conditions such as chemistry, coating condition, and anode performance.

Acknowledgements

This work was supported by the Pipeline Research Council International and by the Gas Research Institute under Contract PR-101-9512.

References

[1] Gas facts: Distribution and transmission miles of pipeline. Technical report, American Gas Association, 1999. www.aga.org/StatsStudies/GasFacts/2131.html.
[2] Pipline accident summary report: Pipeline rupture, liquid butane release, and fire, Lively, Texas August 24, 1996. Technical report, National Transportation Safety Board, Washington, D.C., 1998.
[3] Morgan, J., *Cathodic Protection*. NACE, International: Houston, TX, 2nd edition, 1993.
[4] Wagner, J., *Cathodic Protection Design I*. NACE, International, Houston, TX, 1994.
[5] Peabody, A.W., *Control of Pipeline Corrosion*. NACE: Houston, TX, 1978.
[6] Newman, J.S., *Electrochemical Engineering*. Prentice-Hall: Englewood Cliffs, New Jersey, 2nd edition, 1991.
[7] Orazem, M.E., Esteban, J.M., Kenelley, K.J. & Degerstedt, R.M., Mathematical models for cathodic protection of an underground pipeline with coating holidays: Part 1. theoretical development. *Corrosion*, **53(4)**, pp. 264–272, 1997.

[8] Yan, J.F., Pakalapati, S.N.R., Nguyen, T.V., White, R.E. & Griffin, R.B., Mathematical modeling of cathodic protection using the boundary element method with a nonlinear polarization curve. *Journal of the Electrochemical Society*, **139(7)**, pp. 1932–1936, 1992.

[9] Kennelley, K.J., Bone, L. & Orazem, M.E., Current and potential distribution on a coated pipeline with holidays: Part 1. model and experimental verification. *Corrosion*, **49(3)**, pp. 199–210, 1993.

[10] Nisançioglu, K., Predicting the time dependence of polarization on cathodically protected steel in seawater. *Corrosion*, **43**, pp. 100–111, 1987.

[11] Nisançioglu, K., Gartland, P.O., Dahl, T. & Sander, E., Role of surface structure and flow rate on the polarization of cathodically protected steel in seawater. *Corrosion*, **43**, pp. 710–718, 1987.

[12] Nisançioglu, K. & Gartland, P.O., Current distribution with dynamic boundary conditions. *I. Chem. Symposium Series No. 112: Conference on Electrochemical Engineering*, Loughborough University of Technology: Loughborough, 1989.

[13] Carson, S.L. & Orazem, M.E., Time-dependent polarization of behavior of pipeline-grade in low ionic strength environments. *Journal of Applied Electrochemistry*, **29**, pp. 703–717, 1999.

[14] Anderko, A. & Young, R.D., Model for corrosion of carbon steel in lithium bromide absorption refrigeration systems. *Corrosion*, **56(5)**, pp. 543–555, 2000.

[15] Anderko, A., MacKenzie, P. & Young, R.D., Computation of rates of general corrosion using electrochemical and thermodynamic models. *Corrosion*, **57(3)**, pp. 202–213, 2001.

[16] Korzhenko, A., Tabellout, M. & Emery, J., Dielectric relaxation properties of the polymer coating during its exposition to water. *Materials Chemistry and Physics*, **65(3)**, pp. 253–260, 2000.

[17] Corfias, C., Pebere, N. & Lacabanne, C., Characterization of a thin protective coating on galvanized steel by electrochemical impedance spectroscopy and a thermostimulated current method. *Corrosion Science*, **41(8)**, pp. 1539–1555, 1999.

[18] Margarit, I. & Mattos, O., About coatings and cathodic protection: Possibilities of impedance as monitoring technique. *Electrochemical Methods in Corrosion Research VI, Parts 1 and 2*, Transtec Publications: Zurich-Eutikon, pp. 279–292, 1998.

[19] Thompson, I. & Campbell, D., Interpreting nyquist responses from defective coatings on steel substrates. *Corrosion Science*, **36(1)**, pp. 187–198, 1994.

[20] Bellucci, F. & Nicodemo, L., Water transport in organic coatings. *Corrosion*, **49(3)**, pp. 235–247, 1993.

[21] Diakow, D., Van Boven, G. & Wilmott, M., Polarization under disbonded coatings: Conventional and pulsed cathodic protection compared. *Materials Performance*, **37(5)**, pp. 17–23, 1998.

[22] Jack, T.R., External corrosion of line pipe: a summary of research activities. *Materials Performance*, **35(3)**, pp. 18–24, 1996.

[23] Beavers, J.A. & Thompson, N.G., Corrosion beneath disbonded pipeline. *Materials Performance*, **36(4)**, pp. 13–19, 1997.

[24] Riemer, D. & Orazem, M., Development of mathematical models for cathodic protection of multiple pipelines in a right of way. *Proceedings of the 1998 International Gas Research Conference*, Gas Research Institute, GRI: Chicago, p. 117, 1998. Paper TSO-19.

[25] Riemer, D.P. & Orazem, M.E., Cathodic protection of multiple pipelines with coating holidays. *Proceedings of the NACE99 Topical Research Symposium: Cathodic Protection: Modeling and Experiment*, ed. M.E. Orazem, NACE, NACE International: Houston, TX, pp. 65–81, 1999.

[26] Jones, D.A., *Principles and Prevention of Corrosion*. Prentice Hall: Upper Saddle River, NJ, 1996.

[27] Carslaw, H.S. & Jaeger, J.C., *Conduction of Heat in Solids*. Oxford University Press: New York, 1959.

[28] Moulton, H.F., Current flow in rectangular conductors. *Proceedings of the London Mathematical Society*, **ser. 2(3)**, pp. 104–110, 1905.

[29] Bowman, F., *Introduction to Elliptic Functions with Applications*. John Wiley and Sons: New York, 1953.

[30] Orazem, M.E. & Newman, J., Primary current distribution and resistance of a slotted-electrode cell. *Journal of the Electrochemical Society*, **131(12)**, pp. 2857–2861, 1984.

[31] Orazem, M.E., Calculation of the electrical resistance of a compact tension specimen for crack-propagation measurements. *Journal of the Electrochemical Society*, **132(9)**, pp. 2071–2076, 1985.

[32] Orazem, M.E. & Ruch, W., An improved analysis of the potential drop method for measuring crack lengths in compact tension specimens. *International Journal of Fracture*, **31**, pp. 245–258, 1986.

[33] Diem, C.B., Newman, B. & Orazem, M.E., The influence of small machining errors on the primary current distribution at a recessed electrode. *Journal of the Electrochemical Society*, **135**, pp. 2524–2530, 1988.

[34] Brebbia, C.A. & Dominguez, J., Boundary element methods for potential problems. *Applied Mathematical Modelling*, **1(7)**, pp. 371–378, 1977.

[35] Aoki, S., Kishimoto, K. & Sakata, M., Boundary element analysis of galvanic corrosion. *Boundary Elements VII*, eds. C.A. Brebbia & G. Maier, Springer-Verlag: Heidelberg, volume 1, pp. 73–83, 1985.

[36] Telles, J., Wrobel, L., Mansur, W. & Azevedo, J., Boundary elements for cathodic protection problems. *Boundary Elements VII*, eds. C.A. Brebbia & G. Maier, Springer-Verlag, volume 1, pp. 63–71, 1985.

[37] Zamani, N. & Chuang, J., Optimal-control of current in a cathodic protection system: a numerical investigation. *Optimal Control Applications & Methods*, **8(4)**, pp. 339–350, 1987.

[38] Brichau, F. & Deconinck, J., Numerical model for cathodic protection of buried pipes. *Corrosion*, **50(1)**, pp. 39–49, 1994.

[39] Brichau, F., Deconinck, J. & Driesens, T., Modeling of underground cathodic protection stray currents. *Corrosion*, **52**, pp. 480–488, 1996.

[40] Aoki, S. & Amaya, K., Optimization of cathodic protection system by BEM. *Engineering Analysis with Boundary Elements*, **19(2)**, pp. 147–156, 1997.

[41] Aoki, S., Amaya, K. & Miyasaka, M., Boundary element analysis of cathodic protection for complicated structures. *Proceedings of the NACE99 Topical Research Symposium: Cathodic Protection, Modeling and Experiment*, ed. M.E. Orazem, NACE, NACE International: Houston, TX, pp. 45–65, 1999.

[42] Brebbia, C. & Dominguez, J., *Boundary Elements: An Introductory Course*. McGraw-Hill, The Bath Press: Avon, Great Britain, 1989.

[43] Brebbia, C.A., Telles, J.C.F. & Wrobel, L.C., *Boundary Element Techniques*. Springer-Verlag: Heidelberg, 1984.

[44] Hartmann, F., Katz, C. & Protopsaltis, B., Boundary elements and symmetry. *Ingenieur Archiv*, **55(6)**, pp. 440–449, 1985.

[45] Gray, L. & Paulino, G., Symmetric Galerkin boundary integral formulation for interface and multi-zone problems. *International Journal for Numerical Methods in Engineering*, **40**, pp. 3085–3101, 1997.

[46] Stakgold, I., *Green's Functions and Boundary Value Problems*. John Wiley & Sons: New York, 1979.

[47] Dwight, H.B., Calculations of resistance to ground. *Electrical Engineering*, **55**, p. 1319, 1936.

[48] Telles, J.C.F. & Paula, F.A.D., Boundary elements with equilibrium satisfaction: A consistent formulation for potential and elastostatic analysis. *International Journal of Numerical Methods in Engineering*, **32**, pp. 609–621, 1991.

[49] Trevelyan, J. & Hack, H., Analysis of stray current corrosion problems using the boundary element method. *Boundary Element Technology IX*, Computational Mechanics Publications: Boston, pp. 347–356, 1994.

[50] Burton, S., Garbow, K.E., Hillstrom, J.J. & Others, Modified powell method with analytic jacobian. Software source code hybrdj.f, 1980. Argonne National Laboratory. minpack project. www.netlib.org/minpack.

Boundary Element Technology XV

Editors: *C.A. BREBBIA, Wessex Institute of Technology, UK* and *R.E. DIPPERY, Kettering University, USA*

Numerous special purpose Boundary Element Method (BEM) programmes now exist for a variety of important engineering problems. Containing papers from the Fifteenth International Conference on Boundary Element Technology (BETECH), this book presents some of the most interesting developments in the method. The contributions are divided under headings including: Acoustics; Electrochemical and Corrosion Modelling; Fluid Flow; Fracture and Damage; Heat Transfer; and Stress Analysis.

Series: *Boundary Elements, Vol 4*
ISBN: 1-85312-971-2   2003   408pp
£133.00/US$205.00/€199.50

Boundary Elements XXVII
Incorporating Electrical Engineering and Electromagnetics

Editors: *A. KASSAB, University of Central Florida, USA, C.A. BREBBIA, Wessex Institute of Technology, UK, E. DIVO, University of Central Florida, USA* and *D. POLJAK, University of Split, Croatia*

This book contains the edited proceedings of the 27th World Conference on Boundary Elements together with papers presented at the associated International Seminar on Computational Methods in Electrical Engineering and Electromagnetics.
The Boundary Element Conference series continues to attract original contributions on theoretical and fundamental developments, as well as innovative applications. Its scope has also recently been expanded to include other mesh reduction methods.
The presentations from the Computational Methods in Electrical Engineering and Electromagnetics Seminar cover a wide variety of theoretical and applied topics.
Over 65 papers are included and these are divided under the following headings: BOUNDARY ELEMENTS AND OTHER MESH REDUCTION METHODS - Meshless Methods; Dual Reciprocity Method; Advanced Formulations; Inverse Problems; Stress Analysis; Plates and Shells; Damage Mechanics; Wave Propagation; Fluid Problems; Electrostatics and Electromagnetics; Computational Problems. ELECTRICAL ENGINEERING AND ELECTROMAGNETICS – Interaction of Humans with Electromagnetic Fields; High Frequency Electromagnetic Field Coupling to Transmission Lines; Numerical and Computational Methods; Electrical Engineering and Electronics.

Series: *Advances in Boundary Elements, Vol 20*
ISBN: 1-84564-005-5   2005   760pp

WIT Press is a major publisher of engineering research. The company prides itself on producing books by leading researchers and scientists at the cutting edge of their specialities, thus enabling readers to remain at the forefront of scientific developments. Our list presently includes monographs, edited volumes, books on disk, and software in areas such as: Acoustics, Advanced Computing, Architecture and Structures, Biomedicine, Boundary Elements, Earthquake Engineering, Environmental Engineering, Fluid Mechanics, Fracture Mechanics, Heat Transfer, Marine and Offshore Engineering and Transport Engineering.

Boundary Element Methods for Electrical Engineers

Editors: **D. POLJAK**, *University of Split, Croatia and* **C.A. BREBBIA**, *Wessex Institute of Technology, UK*

This is the first text on Boundary Element Methods specifically for electrical engineers. Written in the form of a primer, it presents the BEM in a simple fashion that will help beginners understand its very basic principles. The authors begin by deriving the BEM for the simplest potential problems, before building on these to formulate methods for a wide range of applications in electromagnetics.

Introducing boundary element fundamentals in a way which will enable readers to solve complex problems on their own, the book is designed for undergraduate and graduate students of electronics and/or electrical engineering. It will also be useful to researchers and professional engineers who wish to exploit the full potential of the BEM in electrical engineering.

Series: Advances in Electrical Engineering and Electromagnetics, Vol 4
ISBN: 1-84564-033-0   2005   240pp
£79.00/US$126.00/€118.50

We are now able to supply you with details of new WIT Press titles via E-Mail. To subscribe to this free service, or for information on any of our titles, please contact the Marketing Department, WIT Press, Ashurst Lodge, Ashurst, Southampton, SO40 7AA, UK
Tel: +44 (0) 238 029 3223
Fax: +44 (0) 238 029 2853
E-mail: marketing@witpress.com

Viscous Incompressible Flow
For Low Reynolds Numbers

M. KOHR and **I. POP**, *Babes-Bolyai University, Cluj-Napoca, Romania*

This book presents the fundamental mathematical theory of, and reviews state-of-the-art advances in, low Reynolds number viscous incompressible flow. The authors devote much of the text to the development of boundary integral methods for slow viscous flow pointing out new and important results. Problems are proposed throughout, while every chapter contains a large list of references.

A valuable contribution to the field, the book is designed for research mathematicians in pure and applied mathematics and graduate students in viscous fluid mechanics.

Contents: Introduction; Fundamentals of Low Reynolds Number Viscous Incompressible Flow; The Singularity Method for Low Reynolds Number Viscous Incompressible Flows; The Theory of Hydrodynamic Potentials with Application to Low Reynolds Number Viscous Incompressible Flows; Boundary Integral Methods for Steady and Unsteady Stokes Flows; Boundary Integral Formulations for Linearized Viscous Flows in the Presence of Interfaces; List of Symbols; Index.

Series: Advances in Boundary Elements, Vol 16
ISBN: 1-85312-991-7   2004   448pp
£148.00/US$237.00/€222.00

All prices correct at time of going to press but subject to change.
WIT Press books are available through your bookseller or direct from the publisher.